Antony Flew was born in London in 1923. During the Second World War he worked in RAF Intelligence. He studied at the School of Oriental and African Studies, London and at Oxford. After lecturing for some years, he was Professor of Philosophy at Keele from 1954 to 1971 and at Reading from 1973 to 1982. He is now Professor Emeritus, University of Reading and Professor of Philosophy at York University, Toronto. Antony Flew has been a Vice-President of the Rationalist Press Association since 1976. He has published many philosophical works, including *Evolutionary Ethics* (1967), *An Introduction to Western Philosophy* (1971), *Sociology, Equality and Education* (1976) and *A Rational Animal* (1978), and has edited a number of volumes, including *Logic and Language* (1951 and 1953) and *Malthus: An Essay on the Principle of Population* (1971).

ANTONY FLEW

DARWINIAN EVOLUTION

SERIES EDITOR
JUSTIN WINTLE

Granada Publishing

Paladin Books
Granada Publishing Ltd
8 Grafton Street, London W1X 3LA

A Paladin Paperback Original 1984

ISBN 0-586-08442-8

Printed and bound in Great Britain by
Hazell Watson & Viney Ltd, Aylesbury

Set in Baskerville

ACKNOWLEDGEMENTS

My thanks are due and most gladly given to the staffs of the libraries of York University, Toronto, and of the University of Reading, England; also to Miss Joan Morris who transformed a legible but unlovely manuscript into a typescript which certainly appeared better than it perhaps is.

ANTONY FLEW, Reading

ACKNOWLEDGMENTS

[illegible]

[illegible]

CONTENTS

I

FROM DARWIN'S *ORIGIN* TO TODAY'S SYNTHETIC THEORY

> By the present time, one can be certain that any theory which successfully explains the entire Past will inevitably – through that achievement alone – win the intellectual leadership of the future.
>
> Auguste Comte, *Discours sur l'Esprit Positif* (1844)

The scope and the limitations of the present book are indicated both by the title, *Darwinian Evolution*, and by its inclusion in a series treating Movements and Ideas. The key word is 'evolution', and the starting point is Darwin. So Chapter I has to begin with an account of Darwin's own life and work, followed by a summary of his theory as it was first publicly presented in *The Origin of Species*. From a consideration of the difficulties and the objections raised there and in the following years we shall proceed to see how these difficulties and these objections have in large part been overcome in the progress of the research and by the development of the modern Synthetic Theory. (The synthesis referred to consists in incorporating a genetic account of variation into the traditional Darwinian structure.)

Darwinian Evolution, however, is not offered as a treatise on evolutionary biology: such a treatise would have to be written, as of course such treatises have been written, by a biologist rather than by a philosopher. Instead, in a book in a series about Movements and Ideas, a biographical and biological first chapter is no more than an essential preliminary. The proper and distinctive business begins only in Chapter II, with an examination of the deductive core of Darwinism, and a general discussion of its actual and alleged philosophical presuppositions and philosophical implications. The third chapter deals with three major and more particular Movements and Ideas topics: the relations and lack of relations between the work of Darwin and that of such social science predecessors as Malthus and the Scottish founding fathers; the Marxist contention that

Marx did for sociology what Darwin did for biology; and the present claims and possible scope of the projected future science of sociobiology. Finally, the fourth chapter examines the suggestions that Darwin's theory provides a guarantee of progress and a foundation for an evolutionary ethics. Although both these suggestions have to be dismissed, we both can and should see many of the greatest issues of our time in the perspective provided by the development of that theory.

1. CHARLES DARWIN (1809–82)

He was born in Shrewsbury, the second son and fifth of the six children of Dr Robert Darwin, a successful and somewhat formidable physician. His mother, who died when Charles was only eight years old, was born Susannah Wedgwood, daughter of Josiah Wedgwood the founder of the famous and still flourishing Staffordshire pottery firm. Robert Darwin's father Erasmus also practised as a doctor, but was better known as the author of *Zoönomia, or the Laws of Organic Life*. Published in 1794 this argued for the mutability of species: Erasmus Darwin recognized the importance of many of the phenomena which were only later and in the light of his grandson's work to become more fully understood. He may, therefore, be accounted an early believer in the evolution as opposed to the special creation of species. He remained, nevertheless, entirely at a loss to suggest any mechanism through which such evolution might have or must have come about. Three years before the grandson died he was to produce, as an exercise in biohistorical scholarship and familial piety, *The Life of Erasmus Darwin*.

Until the spring of 1817, when he was sent to a day school run by the local Unitarian minister, Charles Darwin's elder sister Caroline did her best to teach him at home. In the summer of 1818 he entered Shrewsbury School, but, although it was very close to the family home, as a boarder. He was in his prescribed school work quite undistinguished. He seems, however, to have done a well above average amount of serious reading, while his enthusiasm for natural history, as well as for some other less intellectual country pursuits, certainly began very young. It once provoked his admired and beloved father into the memorably misguided paternal judgement: 'You care for nothing but shooting, dogs, and rat-catching, and you will be a disgrace to yourself and all your family' (Barlow, p. 28).

In the autumn of 1825, at what was even in those days the unusually young age of sixteen, Charles was sent to Edinburgh University. He followed in the steps of his elder brother Erasmus, who was due to graduate in medicine at the end of the academic year 1825–6. Partly because he had got wind of the fact that he could expect an inheritance sufficient to enable him to live without any professional practice, partly because the boredom of the lectures in the medical school was at that period excruciating, and partly because he was repelled by the spectacle of bloody tortures in surgical operations performed without anaesthetics, Charles left Edinburgh after two years.

Not without a touch of cynicism, his unbelieving father Dr Robert Darwin then suggested that his second son should transfer to Christ's College, Cambridge, there to train for admission into the priesthood of 'the Church by law established'. Charles, as he was later to confess in his *Autobiography*, 'had scruples about declaring my belief in all the dogmas of the Church of England'. But these were soon put at rest: 'I read with care *Pearson on the Creed* and a few other books on divinity; and, as I did not then doubt the strict and literal truth of every word in the *Bible*, I soon persuaded myself that our Creed must be fully accepted. It never struck me how illogical it was to say that I believed in what I could not understand and what is in fact unintelligible' (ibid., p. 57).

All doubts thus resolved, Charles next discovered that he had forgotten the Latin and Greek learnt at Shrewsbury School. So the first term of 1827–8 had to be spent cramming with a private tutor. He went into residence only in January 1828. He was later to write: 'During the three years I spent at Cambridge my time was wasted, as far as the academical studies were concerned, as completely as at Edinburgh and at school . . . my time was sadly wasted there and worse than wasted. From my passion for shooting and for hunting and, when this failed, for riding across country I got into a sporting set, including some dissipated low-minded young men' (ibid., pp. 58 and 60).

There was only one part of the work prescribed by his tutors which Darwin thought worthwhile: 'In order to pass the BA examination, it was . . . necessary to get up Paley's *Evidences of Christianity* and his *Moral Philosophy*. This was done in a thorough manner, and I am convinced that I could have written out the whole of the *Evidences* with perfect correctness . . . The logic of this book and, as I may add, of his *Natural Theology* gave me as much delight as did Euclid' (ibid., p. 59).

The importance of this training in logical thought is not to be minimized. Darwin was later to describe *The Origin of Species* itself as 'one long argument from beginning to end'. Nor should we overlook, as Darwin seems here to have done, the significance of the fact that his favourite models of rigorous thought were provided by Paley. For Paley, especially in his *Natural Theology*, offers as evidence of divine design case after case of astonishing biological adaptation. He was also among the first and most distinguished converts to be won by Thomas Malthus, whose *Essay on Population* gave an indispensable stimulus to the construction of Darwin's theory of evolution by natural selection.

But, in addition to these perhaps none too strenuous studies for a not altogether discreditable Pass BA, Darwin was still reading widely and carefully. He was also continuing his biological and geological education. Already in Edinburgh he had made modest discoveries: 'that the so-called ova of Flustra had the power of independent movement by means of cilia, and were in fact larvae'; and 'that little globular bodies which had been supposed to be the young state of *Fucus loreus* were the egg-cases of the worm-like *Pontobdella muricata*' (ibid., p. 50). In Darwin's Cambridge there were no degree courses in any non-mathematical sciences. But the chairs in mineralogy, geology and botany were held, respectively, by William Whewell, Adam Sedgwick and John Stevens Henslow; all three considerable men. Darwin early formed a lifelong friendship with Henslow, and through him became acquainted with the other two. Darwin thus gained an informal education in these sciences in a way which would scarcely be possible for a modern undergraduate.

To appreciate how much this meant for him it is worth pondering for a moment his own statement in the *Autobiography*: 'I have not as yet mentioned a circumstance which influenced my whole career more than any other. This was my friendship with Prof. Henslow . . . He kept open house once every week, where all undergraduates and several older members of the University, who were attached to science, used to meet in the evening . . . Before long I became well acquainted with Henslow, and during the latter half of my time at Cambridge took long walks with him on most days; so that I was called by some of the dons "the man who walks with Henslow"; and in the evening I was very often asked to join his family dinner' (ibid., p. 64).

After passing his examinations in January 1831 – coming tenth on the list of those not aspiring to honours – Darwin still had two more terms of residence to put in. On Henslow's advice he devoted himself to geology, going back on his Edinburgh resolve never again to touch that subject. He had also recently read Alexander von Humbolt's *Personal Narrative of Travels* and John Herschel's introduction to *The Study of Natural Philosophy*, from both of which he derived such a zeal to travel and to study natural history that he started to plan a journey to Tenerife, inquired about ships, and learnt Spanish. On his return home at the end of August from a geological excursion to North Wales in the company of Adam Sedgwick he found letters waiting for him with an invitation to join Captain FitzRoy of HMS *Beagle* on a journey round the world, as a naturalist and captain's companion, without pay.

Henslow himself, under pressure from his wife, had already refused this invitation. His letter to his younger friend and protégé merits quotation:

I have stated that I consider you to be the best qualified person I know of who is likely to undertake such a situation: I state this not in the supposition of your being a *finished* naturalist, but as amply qualified for collecting, observing and noting, anything worthy to be noted in Natural History . . . Don't put any modest doubts or fears about your disqualifications, for I assure you I think you are the very man they are in search of. [Here as elsewhere, except where notice is given, the emphasis is in the original.]

After a strenuous intervention by his uncle Josiah Wedgwood, overcoming Dr Robert Darwin's anxieties about the implications of this project for a future career as a clergyman, Charles Darwin accepted Henslow's excellent advice.

So, after a frenzy of preparations and equally frenetic delays, the tiny but unusually well equipped *Beagle* set sail from Plymouth on 27 December, not to return to England until nearly five years later. Darwin described the voyage of the *Beagle* as 'by far the most important event' in his life, one which 'has determined my whole career' (ibid., p. 76). It makes a splendid story of travel, observation, speculation and adventure. Although tormented by seasickness, and despite living in preposterously close and confined conditions, he got through an unbelievable amount of scientific work. Every opportunity to explore ashore was seized. The thirty-mile-a-day walking tours in North Wales paid off. Nor had those long cross-country rides with 'a sporting set, including some dissipated low-minded young men' gone for nothing. If a mountain was there, he had to climb it. If he had a chance of a 500-mile ride across the pampas, living as a gaucho, he took it. Nor were his observations exclusively natural as opposed to social. For instance: confronted in Brazil by the institution of slavery he was, like his parents and grandparents before him, outraged. (His enterpriser grandfather was an early campaigner for emancipation, who produced propaganda medallions asking on behalf of a black slave: 'Am I not a man and a brother?')

Again, one of his long rides led him through the armies of the future Argentine dictator General Rosas, then in command of a savage war of extermination against the indigenous Indians. (Contemporary British readers, having had recent experience of Argentine colonialism, will also note Darwin's remarks on the then recent annexation of the almost uninhabited East Falklands – 'a wretched place'.)

Immediately on his return, Darwin set about arranging his collections for exhibition and his findings for publication. He became in this period accepted as a leading professional scientist, although he was never to hold any paid position. There had for a long time been no further talk of his taking Holy Orders. In November 1838 he proposed to and was accepted by his cousin Emma Wedgwood. In January 1839 they married, five days after his election to a Fellowship of the Royal Society.

Later in that year there appeared his *Journal of Researches into the Natural History and Geology of the Countries Visited during the Voyage of HMS 'Beagle' Round the World, under the Command of Captain Fitzroy R.N.*, a work nowadays known by the less spaciously Victorian title *The Voyage of the 'Beagle'*. It thoroughly deserved and has in fact continued to be widely read. (It is at the time of writing available both in hard-cover and in paperback in the Everyman Library. There is also a very well written and illustrated account of *Darwin and the Beagle*, based on a filmscript for BBC television, by Alan Moorehead.)

Darwin had in July 1837 opened his first 'Notebook on the Transmutation of Species', although thoughts on this subject had been brewing for some time, especially after *Beagle*'s visit to the Galapagos Islands in September 1835. How could these several islands have come to support different species of finch and turtle, all adapted to fill particular ecological niches on their own particular islands, if they had really had no common ancestry among incomers from the distant yet not impossibly remote South American mainland?

In September 1838 a reading of Malthus on population – the *Second Essay* – provided the crucial stimulus. How did those various finches and turtles become so perfectly adapted to their several particular ecological niches? Because, under the constant, crushing pressure of population, any variant forms possessing some competitive advantage over the normals must tend to survive and to reproduce, while the (former) normals must tend not to reproduce and to die out.

In 1842, after the birth of their first two children and the publication of a major geological contribution, *The Structure and Distribution of Coral Reefs*, the Darwins moved to the seclusion of Down House in Kent, now preserved as a museum. Already Charles had begun to suffer from the spells of pain and exhaustion which were to plague him for the rest of his life. Partly because his most learned medical contemporaries were unable to present any satisfactory diagnosis of his condition, partly because we all find it hard to believe a chronic invalid could have accomplished so much, and partly because a psychologically over-sophisticated age refuses to believe that any character could be so open and ingenuous, it is often suggested that he was a hypochondriac or, if there is any difference behind this distinction, that the source of his trouble was psychosomatic.

There is, I believe, no longer any warrant for such suspicions. So we may with a clear conscience accept his future wife's testimony: 'He is the most open transparent man I ever saw, and every word expresses his real thoughts.' For medical research has since identified Chagas's Disease; and its symptoms seem to correspond with all the case reports on Charles Darwin. (Since neither the Darwin nor the Wedgwood families seem ever to have thrown away any written papers, such reports are in one form or another abundant!) And, furthermore, we know from Darwin's *Beagle* journals that on 26 March 1835 he was extensively bitten by *Triotoma infestans*, the nasty creature which carries this even nastier disease.

By the end of June 1844 Darwin had completed what he called 'a sketch of my species theory'. He then wrote in a letter to his wife, dated 5 July, that 'If, as I believe, my theory in time be accepted by one competent judge, it will be a considerable step in science. I therefore write this in case of my sudden death, as my most solemn and last request, which I am sure that you will consider the same as if legally entered in my will, that you will devote £400 to its publication, and further, will yourself, or through Hensleigh [Mr H. Wedgwood], take trouble in promoting it' (Darwin, F., II p.16). So the question arises why Darwin, who clearly realized what a treasure he had in his hands, did not push ahead to immediate publication? The answer, surely, is that he knew that publication would stir a frightful scandal; that he was in temperament a cautious man; and that he wanted to wait until either the climate of opinion changed and/or he had accumulated quite overwhelming evidential support for his speculations.

Certain it is that in October 1846, after finishing one or two minor jobs, he launched out into an enormous and teutonically thorough study of barnacles. This was to occupy him for the next ten years. Thereby hangs a tale, worth telling in order to hint at the happiness and devotion of family life in Down House. In this period one of his sons was out to tea with some small friends when he noticed something missing in the home of his hosts. So he asked of their father: 'But *where* does he do his barnacles?' On another occasion Darwin's children teased him because his enthusiastic descriptions of species read like advertisements. The larval cirripede, for instance, is endowed 'with six pairs of beautifully constructed legs, a pair of magnificent compound eyes, and extremely complex antennae'.

In May 1856 he began work on what was to be an immense treatise on species. But in June 1858 he received from Alfred Russel Wallace a paper outlining his own treasured theory, a theory which Wallace under the same Malthusian stimulus had developed much later but quite independently. Darwin was at first shattered, though from the beginning resolved to react with absolute propriety. Fortunately, at the urging of his great friends the geologist Sir Charles Lyell and the naturalist J. D. Hooker, Darwin agreed to allow an extract from his manuscript, plus a letter to his American confidant the botanist Asa Gray, to be published in the *Journal of the Proceedings of the Linnaean Society* along with Wallace's essay 'On the Tendency of Varieties to depart indefinitely from the Original Type'. At the end of that year the society's president concluded in his annual address that it had not been 'marked by any of those striking discoveries which at once revolutionize . . . the department of science on which they bear'!

Darwin, on the further urging of the same two friends, began in September 1858 to abstract from what he already had of the projected massive treatise and from other manuscripts what became his still moderately massive *The Origin of Species*. This certainly did not, as David Hume said of his own *A Treatise of Human Nature*, 'fall dead-born from the press'. The first edition of 1,250 copies sold out on the day of publication, 24 November 1859, the second of 3,000 soon afterwards, and the book within months had received at least ninety reviews and critical notices.

Four further editions followed, all requiring extensive preparatory revision, as well as several other books. In 1868 there was the *Variation of Plants and Animals under Domestication* and in 1872 *The Expression of the Emotions in Man and Animals*. Towards the end of the final chapter of *The Origin of Species* Darwin treated himself and his readers to an exquisite and discreet masterpiece of understatement: 'In the distant future I see open fields for far more important researches . . . Light will be thrown on the origin of man and his history' (Darwin, C., 1859, p. 458). In the event, that future proved to be not distant at all. For in 1871 – against the background of the flames of the Paris Commune, as one reviewer complained – Darwin published *The Descent of Man, and Selection in Relation to Sex*.

In 1881, the year before his death, he called on his publisher carrying a final manuscript: 'Here is a work which has occupied me for many years. and interested me much. I fear the subject will not interest the public, but will you publish it for me?' Darwin's fears were baseless. *The Formation of Vegetable Mould through the Action of Worms* was published in October 1881, and six printings were called for in less than a year. The author died at Down House in April 1882. Although the family wanted to bury him in the village churchyard, on the initiative of twenty members of both houses of Parliament, a body at that time including four Fellows of the Royal Society, the Dean of Westminster agreed to an Abbey burial. This was, wryly, the only honour which this very great Englishman ever received from official Britain. He rests now close to Newton, Faraday and his friend Lyell.

2. THE ARGUMENT OF *THE ORIGIN OF SPECIES*

Thomas Henry Huxley, staunch friend and supporter of Charles Darwin, several times reviewed *The Origin of Species*: both on its first publication; and then again more than twenty-five years later, after the author's death. In *The Times* for 26 December 1859 Huxley started, as did Darwin, with 'Variation under Domestication', pointing out that the (artificial) breeders of plants, pigeons, dogs and cattle all consider themselves to have produced a variety of fresh species.

> But [he goes on] in all these cases we have human interference. Without the breeder there would be no selection . . . Before admitting the possibility of natural species having originated in any similar way, it must be proved that there is in nature some power which takes the place of man . . . It is the claim of Mr Darwin . . . to have discovered the existence and the modus operandi of this natural selection, as he terms it; and, if he be right, the process is perfectly simple and comprehensible, and irresistibly deducible from very familiar but well nigh forgotten facts. (Huxley, T. H., 1859, p. 348)

Then in one piece, first written in 1887, on 'The Progress of Science' Huxley urged that 'So far as biology is concerned, the publication of *The Origin of Species*, for the first time, put the doctrine of evolution, in its application to living things, upon a sound scientific foundation. It became an instrument of investigation, and in no hands did it prove more brilliantly profitable than in those of Darwin himself' (Huxley, T. H., 1887, p. 101). Again in the same year, in a piece 'On the Reception of *The Origin of Species*', he concluded: 'And even a cursory glance at the history of the biological sciences during the last quarter of a century is sufficient to justify the assertion, that the most potent instrument for the extension of the realm of natural knowledge . . . since the publication of Newton's *Principia*, is Darwin's *Origin of Species*' (Darwin, F., II p. 204).

(a) His own apt and carefully chosen title is rarely given in full. It is *On the Origin of Species by Means of Natural Selection*. There was also a subtitle, which has since, for all who were alive and alert in the late 1930s and early '40s, acquired a sinister ring: *or The Preservation of Favoured Races in the Struggle for Life*. Darwin's theory is evolutionary inasmuch as it asserts, and provides an account of, the developmental rather than catastrophic origin of species. Evolution is here to be contrasted with the previously dominant notion of the fixity of species, all of which had presumably been specially created in substantially their present form.

That is assumed, for instance, indeed asserted, in the picturesque creation stories of Genesis 1 and 2:

> And God said: Let the earth bring forth the living creature after his kind, cattle, and creeping thing, and beast of the earth after his kind. And it was so. And God made the beast of the earth after his kind, and the cattle after their kind, and everything that creepeth upon the ground after his kind; and God saw that it was good. (1: 24–5)

The contrary doctrine is, thus, that the first specimens of every species were individually created. It is from these first parents – normally if not perhaps quite always breeding true – that the earth has been filled. Every species is, to use a Greek rather than a purely Hebrew notion, a natural kind. No member of any such natural kind could ever, in the ordinary course of nature, evolve or develop into a specimen of anything else. Classification, too, becomes a matter of discovering to what pre-existing plans the creatures to be classified do in fact conform. To these ideas the Genesis story of Adam naming the beasts is altogether appropriate: 'And out of the ground the Lord God formed every beast of the field, and every fowl of the air; and brought them unto Adam to see what he would call them: and whatsoever Adam called every living creature, that was the name thereof' (2: 18). The same ideas guided, or misguided, the Swedish Academy when it honoured the great eighteenth-century systematist (taxonomist) Linnaeus: 'He discovered the essential nature of insects.'

In thus presenting an evolutionary as opposed to catastrophist account of the origin of species, Darwin was trying to do for biology what his own senior friend Charles Lyell had already done for geology; and what, we may add, Immanuel Kant had, as early as 1755, attempted to achieve in cosmology with his *General History of Nature and Theory of the Heavens*. The long but informative title of Lyell's main work was *The Principles of Geology, Being an Attempt to explain the Former Changes of the Earth's Surface, by Reference to Causes now in Operation*. The first volume came out just when HMS *Beagle* was about to sail. Darwin took a copy of it with him, had the remainder posted on to him in South America, and was in short order completely converted. Indeed, at that stage he thought of himself as primarily a geologist, and his first published scientific papers were in geology. It is this spread of evolutionary and developmental ideas into area after area which led two leading historians of science, writing on *The Discovery of Time*, to assert: 'The picture of the natural world which we all take for granted today has one remarkable feature, which cannot be ignored in any study of the ancestry of science: it is a *historical* picture' (Toulmin and Goodfield, p. 17; and compare pp. 129–35 on Kant's contribution).

The general idea of evolution, of the mutability of species, was not, as we have seen already in Section 1, original to Darwin. What immediately distinguished his book from such predecessor volumes as his grandfather's *Zoönomia*, or Jean-Baptiste de Lamarck's *Philosophie zoologique* (1809), or from *Vestiges of the Natural History of Creation* (1844) by Robert Chambers, was Darwin's suggestion about the controlling and determining mechanism of biological evolution: *On the Origin of Species by Means of Natural Selection*. It is 'natural selection' which is the crucial notion in an account not of that but of how species must have originated, and must still be originating.

Yet even this was no more entirely new than any of the other notions incorporated into his fundamental conceptual scheme. The originality and the greatness of Darwin's achievement lie elsewhere. He was not, nor of course did he ever claim to be, the first to assert the evolution, as opposed to the special creation, of species: 'The general hypothesis of the derivation of all present species from a small number, or perhaps a single pair, of original ancestors was propounded by the President of the Berlin Academy of Sciences, Maupertuis, in 1745 and 1751, and by the principal editor of the *Encyclopédie*, Diderot, in 1749 and 1754' (Lovejoy, 1936, p. 268; and compare Lovejoy, 1904, passim).

Nor was Darwin the first to introduce into a biological context the ideas either of natural selection or of a struggle for existence. Both can be found in the Roman poet Lucretius in the first century B.C., in an account of how in the beginning our mother earth produced both all the kinds of living things which we now know, and many other sorts of ill-starred monstrosity. But with these latter:

> . . . it was all in vain . . . they could not attain the desired flower of age nor find food nor join by the ways of Venus . . . And many species of animals must have perished at that time, unable by procreation to forge out the chain of posterity; for whatever you see feeding on the breath of life, either cunning or courage or at least quickness must have kept . . . from its earliest existence. (V pp. 845–8 and 855–9)

Lucretius, too, was a disciple, clothing in Latin verse ideas which he had himself learnt from the fourth-century Greek Epicurus, who was here in his turn drawing on such fifth-century sources as Empedocles of Acragas (Kirk and Raven, pp. 336–40).

Darwin himself was well aware of, and indeed took delight in pointing to, many of these partial anticipations – especially, perhaps, those compassed by his own engaging grandfather Erasmus. But attention to such anticipations must not be allowed to distract us from the reasons why in his new employment separately ancient ideas had, and deserved to have, so enormous an impact. These reasons are: first, that Darwin put it all together, assembling the scattered conceptual essentials into a single deductive scheme; and, second, that under the control of that scheme Darwin deployed an immense mass of supporting evidence – much of this the product of his own fieldwork in the years in *Beagle* and after. Not for nothing was Huxley both to speak of 'the publication of that theorem by Darwin and Wallace' (Huxley, T. H., 1887, p. 100), and to excuse his own numerous citations of that supporting evidence in wry words: 'If a man *will* make a book, professing to discuss a single question, an encyclopaedia, I cannot help it' (Huxley, T. H., 1906–, p. 245).

It was precisely this overwhelming argument which slowly convinced the contemporary scientific world that, whatever other factors in the process might remain still to be discovered, evolution by natural selection must have occurred, and be still occurring. Furthermore, overwhelming though it is, *The Origin of Species* – quite unlike *Principia* – can be understood immediately by the intelligent layperson. The combined effect was, and remains, to make it imperative for anyone with any pretensions towards a rational and comprehensive world outlook to try to come to terms with Darwinism.

(b) So how does the argument go? The 'Introduction' indicates, both that the book has a deductive skeleton, and what that skeleton is: 'As many more individuals of each species are born than can possibly survive; and as, consequently, there is a frequently recurring struggle for existence, it follows that any being, if it vary however slightly in any manner profitable to itself, under the complex and sometimes varying conditions of life, will have a better chance of surviving and thus be *naturally selected*. From the strong principle of inheritance, any selected variety will tend to propagate its new and modified form' (Darwin, C., 1859, p. 68). He also promises that in the chapter 'Struggle for Existence' he will treat this struggle 'amongst all organic beings throughout the world, which inevitably follows from their high geometrical powers of increase' (ibid., p. 68; notice what Kant would have loved to call the apodeictic terms and expressions 'consequently', 'it follows that', and 'which inevitably follows from').

But the first clues, providing purchase for the application of this argument, are the facts of Chapter I 'Variation under Domestication'. From this study we 'see that a large amount of hereditary variation is at least possible; and, what is equally or more important, . . . how great is the power of man in accumulating by his Selection successive slight variations'. In all this we see Darwin the countryman born and bred, expressing his 'conviction of the high value of such studies, although they have very commonly been neglected by naturalists' (ibid., p. 67).

More positively charming is the picture called up by his statement to Huxley three days after publication: 'I have found it very important associating with fanciers and breeders. For instance, I sat one evening in a gin palace in the Borough amongst a set of pigeon fanciers, when it was hinted that Mr Bull had crossed his Pouters with Runts to gain size; and if you had seen the solemn, the mysterious, and awful shakes of the head which all the fanciers gave at this proceeding, you would have realized how little crossing has to do with improving breeds' (Darwin, F., II pp. 281–2). Who can fail to relish this cameo of the rich, sober, massively respectable Fellow of the Royal Society, accepted into a set which must have included many 'dissipated and low-minded . . . men', immediately it emerged that his own family bred pigeons and sheep?

Besides relishing a happy social situation, we should note the significance of this appeal to the experience of the pigeon fanciers. For several varieties which are, at least to superficial inspection, quite vastly different certainly have been produced, by artificial selection, within a period which is, in terms of geological time scales, one of days rather than of years. And these have been produced from variations arising among descendants from a common origin, rather than by cross-breeding.

Chapter II continues with 'Variation under Nature'. Chapter III deals with 'Struggle for Existence'. This is both itself an observed fact and a necessary inference from other known facts; the facts, that is, both of the finitude of the earth and of the multiplicative powers of living things. It is perhaps worth noting right now that Darwin never claimed either to have observed or to be able to infer an unrelenting war of all against all. His account of the struggle for existence in the natural world leaves abundant room for *mutual aid* between members both of the same and of different species. As an omnivorous naturalist, he had himself either observed or read of most of the examples cited by Prince Kropotkin. (It is the same in the less lethal world of commerce and industry: even in the most healthily competitive economy there has to be and is co-operation within as well as competition between firms; while several firms may co-operate in forming a consortium in order to compete for a big contract.)

In this Chapter III Darwin develops the argument: 'A struggle for existence inevitably follows from the high rate at which all organic beings tend to increase . . . as more individuals are produced than can possibly survive, there must in every case be a struggle for existence, either one individual with another of the same species, or with the individuals of distinct species, or with the physical conditions of life. It is the doctrine of Malthus applied with manifold force to the whole animal and vegetable kingdom; for in this case there can be no artificial increase of food, and no prudential restraint from marriage. Although some species may be now increasing, more or less rapidly, all cannot do so, for the world would not hold them' (Darwin, C., 1859, pp. 116–17).

Just as the struggle for existence is derived as a consequence of the combination of powers of multiplying indefinitely with always finite resources to support life, so in Chapter IV, 'Natural Selection', this in turn is derived as a consequence of the combination of the struggle for existence with variation. Darwin summarizes his argument here: 'If . . . organic beings vary at all in the several parts of their organization, and I think this cannot be disputed; if there be, owing to the high geometrical powers of increase of each species, at some age, season, or year a severe struggle for life, and this certainly cannot be disputed; then . . . I think it would be a most extraordinary fact if no variation had ever occurred useful to each being's own welfare, in the same manner as so many variations have occurred useful to man. But if variations useful to any organic being do occur, assuredly individuals thus characterised will have the best chance of being preserved in the struggle for life; and from the strong principle of inheritance they will tend to produce offspring similarly characterised' (ibid., pp. 169–70).

(c) However, Darwin continues, 'Whether natural selection has really thus acted in nature . . . must be judged of by the general tenour and balance of evidence given in the following chapters' (ibid., p. 170). But before proceeding further let us notice two in principle falsifiable yet not in fact falsified implications of the theory: '. . . we already see how it entails extinction; and how largely extinction has acted in the world's history, geology plainly declares. Natural selection, also, leads to divergence of character; for more living beings can be supported on the same area the more they diverge in structure, habits, and constitution, of which we see proof by looking at the inhabitants of any small spot' (ibid., p. 170).

Falsifiability is an essential requirement in any truly scientific theory. It is essential because, if a theory is to explain why this or that happens, it must at the same time and by the same token explain why it is this or that, and not anything else, which does in fact happen. The theory which can be reconciled with anything which might conceivably occur or which might conceivably have occurred, with not only what does or did in fact happen but also with what does not or did not yet conceivably might or might have, does not genuinely explain anything at all. It is not a possibly true or possibly false scientific theory. For it is not a scientific theory at all. This fundamental logical insight is, of course, the basis of Sir Karl Popper's whole philosophy of science; something which has had and continues to have the widest and most salutary influence (Popper, 1963; and compare Magee).

Now Darwin here appreciates that his theory carries two major implications, both of which not merely can be but have been tested, and found to be true. It is, therefore, in at least these two respects – as well as, as we shall be seeing later in others – falsifiable but not in fact falsified. So it is perhaps a little hard to see how Popper reached his own contrary conclusion, since retracted: 'I have come to the conclusion that Darwinism is not a testable scientific theory but a *metaphysical research programme* – a possible framework for testable scientific theories' (Popper, 1974, I p. 134). The first of the present two falsifiable implications is shown not to be false by the record of the rocks: '. . . it entails extinction; and how largely extinction has acted in the world's history, geology plainly declares.' Though by then plain this declaration was in Darwin's day still comparatively recent. For, on the basis of their very different presuppositions, both John Wesley and Thomas Jefferson denied both the fact and the possibility of extinction: 'Death', Wesley proclaimed in 1770, 'is never permitted to destroy the most inconsiderable species'; while Jefferson assured us that 'the economy of nature' is such 'that no instance can be produced of her having permitted any one race of animals to become extinct; of her having formed any link in her great work, so weak as to be broken'.

The second of the present two falsifiable implications is, because it presupposes the occurrence of the necessary series of adaptive variations, less strong. But Darwin was in fact not short of evidence of that 'divergence of character' which – given that 'more living beings can be supported on the same area the more they diverge in structure, habits and constitution' – his theory would lead us to expect. How, for instance, could he ever fail to recall those so many kinds of finches in the Galapagos, all clearly descended from some original wind-blown immigrants, and each now adapted to its own peculiar, blindly discovered, ecological niche?

Furthermore, as 'Darwin's bulldog' used to ask, what is the alternative? 'I really believe that the alternative is either Darwinism or nothing, for I do not know of any rational conception or theory of the organic universe which has any scientific position at all beside Mr Darwin's' (Huxley, T. H., 1906, p. 258). Or, as Darwin himself asked in Chapter XIV, 'Recapitulation and Conclusion', can anyone really believe what the non-scientific alternative asks us to believe? These opponents 'seem no more startled at a miraculous act of creation than at an ordinary birth. But do they really believe that at innumerable periods in the Earth's history certain elemental atoms have been commanded suddenly to flash into living tissues? Do they believe that at each supposed act of creation one individual or many were produced? Were all the infinitely numerous kinds of animals and plants created as eggs or seed, or as full grown? and, in the case of mammals, were they created bearing the false marks of nourishment from the mother's womb? (Darwin, C., 1859, p. 454)

The final clause of that challenge is one token of a type notoriously and recklessly sustained by Sir Philip Gosse, the father of Edmund Gosse's *Father and Son*. Sir Philip, a considerable naturalist, two years before *The Origin of Species* published *The Natural History of Creation, or Omphalos*; the final word in that title being, significantly, the Greek for navel. Writing at a time when Lyell's uniformitarian and evolutionary views had long since become the almost universally accepted orthodoxy in geology, but when – incongruously, but with no formal inconsistency – most biologists believed that all or most species were independently created, Gosse insisted, where most of his colleagues chose to forget, that any special creation of any creature which is to be truly a normal member of the species concerned must be a creation at some particular stage in its various cycles. Much the same applies, as he also insisted, to rocks; in particular to fossil-bearing sedimentary rocks.

So if, as he firmly believed, God created the universe in or around 4004 B.C., then all organisms, and indeed everything else, then created would have to bear marks of pasts which they never had. Adam and Eve would have had to have navels, while sedimentary rocks would have had to contain the fossil remains of organisms which had never lived: all, surely, powerful temptations to infidelity?

To the whole-hearted scientific naturalist such consequences are bound to be altogether incredible. No doubt Gosse ought to have seen that he had produced a triumphant reduction to absurdity of an idea incongruous with the whole spirit and method of science. He ought to have seen, that is, that there is and can be no scientific alternative to some theory of evolution; whether or not the whole process is completely explicable in purely Darwinian terms. Yet no one who was prepared to go on accepting the conventional wisdom about the special creation of species – as at the time when *Omphalos* was published most people were – had any right to ridicule Gosse – as they mostly did – when with learning and candour he presented its preposterous consequences. Such consequences arise, as he pointed out, for things organic and inorganic both. All these consequences he himself honestly and boldly proclaimed for true, true precisely and only because they were in their turn clear consequences of other and more fundamental propositions, propositions which he and most of them asserted to be true. *Omphalos* is, of course, the book of a pitifully deluded fanatic. Yet it is neither dishonest nor time-serving, neither muddle-headed nor evasive. Gosse himself deserves some credit. Most of his contemporary critics merit none.

3. BLENDING VARIATION AND/OR SALTATORY MUTATION?

In Chapter V, 'Laws of Variation', we reach one of the main areas of weakness. As always, Darwin is forthright and ingenuous: 'Our ignorance of the laws of variation is profound' (Darwin, 1859, p. 202). What, and indeed all that, he feels able to assert with full confidence is: first, that hereditable variations are 'common and multiform in organic beings under domestication, and in a lesser degree in those in a state of nature' (ibid., p. 173); and, second, that 'Whatever the cause may be of each slight difference in the offspring from their parents – and a cause for each must exist – it is the steady accumulation, through natural selection, of such differences, when beneficial to the individual, that gives rise to all the more important modifications of structure, by which the innumerable beings on the face of this earth are enabled to struggle with each other, and the best adapted to survive' (ibid., pp. 203–4).

That is for Darwin's theory fully sufficient, although for a full and sufficient theory we need also to know what the laws of variation actually are. He was himself prepared to allow at least some measure of truth to Lamarck's since totally discredited suggestion that characteristics acquired by use may be inherited. This Huxley subjected to typically robust ridicule in his review in *The Times*. Could all our herons and long-necked waders really be descended from short-necked birds who had persevered in the attempt to catch fish without wetting themselves? Surely any such poorly endowed creature would have 'renounced fish dinners long before it had produced the least effect on leg or neck?' (Huxley, T. H., 1859, p. 345)

More serious difficulties were generated both by Darwin's acceptance of the doctrine of blending inheritance and by his insistence upon 'the canon "Natura non facit saltum"' (Nature does not take leaps). The former maintains that the offspring of a mating of large and small will be just middling. As for the latter it all depends, of course, how big a jump counts as a leap.

The difficulty arising from blending inheritance is that, on this assumption and in a large population, single variations would be bound within a few generations to be overwhelmed; with a consequent asymptotic approach of the entire population to the previous average level in whatever was the relevant respect. This difficulty Darwin laboured to meet by postulating relatively small, isolated populations. The difficulty arising from the traditional canon is, in a nutshell, that, all other things being equal, the shorter the steps taken the longer it must take to cover any given distance. One question which both is and ought to be pressed on any form of Darwinism is: 'Has there been enough time?' So, the longer the leaps for which room can be made in the theory, the better the chances of returning a satisfactory affirmative answer to this challenge.

Commitment to the aforementioned traditional canon is one of the few counts upon which Huxley faulted Darwin. Consistently with this rejection he also refused to accept Darwin's own later hypothesis supposedly accounting for the phenomena of heredity, pangenesis. Already, in their day, there was knowledge of some fairly substantial variations, substantial variations which would lead their twentieth-century successors to speak of the occurrence of genetic mutations. Indeed, Linnaeus as early as the 1740s had studied a quite fresh form of toadflax, which had apparently originated from the normal plant as a true-breeding change of species, or (as he himself put it) as a mutation. Again, the first short-legged Ancona seems to have been born in Massachusetts in 1791 to become the ancestor of a whole new breed of sheep. Such long leaps were called saltations, while any account providing for them used to be described as saltatory (Huxley, T. H., 1860).

It is one of the oddities of the history of science that Gregor Mendel in his Moravian monastery began the work which indicated how these two difficulties about variation could be overcome, a few months after Darwin had started to compose the massive treatise, of which *The Origin of Species* was in the event offered as an abstract. Mendel possessed a copy of the third German edition of that abstract, published in 1863, and in it he was at pains to mark passages dealing with variation and the origin of hybrids. Although he does not mention Darwin in the paper which he read to the Natural History Society of Brno in 1863, and published in 1865, and although as a hard-line Augustinian abbot he would not accept the mutability of species, it is clear that he had the difficulties of Chapter V in mind, and saw how his own findings could go a long way towards meeting them.

Mendel, unfortunately, did not draw his paper to Darwin's attention. So it went unnoticed till 1900, when it was discovered by three men who had independently performed breeding experiments with the same results as those reported from Mendel's sweet peas. They were Carl Correns in Germany, Eric von Tschermak in Austria and Hugo de Vries in Holland. De Vries, working on the evening primrose, then went on to propound a mutation theory, attributing the origin of species to sudden discontinuous variation, and claiming that small blending variations were not the materials for the formation of new species.

As this theory was developed, it appeared to be incompatible with any Darwinian account of evolution by natural selection, albeit for reasons which need not concern us here. But in 1926 S. S. Chetverikov suggested that this might not after all be so. Using the subsequent work by Chetverikov and his colleagues on wild populations of the fruit-fly Drosophila, R. A. Fisher in 1930 succeeded in integrating Darwinian selection and Mendelian genetics into a synthetic theory, by showing that and how, far from being antagonistic, they are mutually indispensable and between them provide exactly the mechanism needed for evolution by natural selection. Through the subsequent progress of the research the definite has come to displace the indefinite article: it is now either Neo-Darwinism or The Synthetic Theory of Evolution.

4. DIFFICULTIES, OBJECTIONS AND SOME ANSWERS

Chapter VI is entitled 'Difficulties on Theory'. But other, subsequent chapters also deal with difficulties and the overcoming of difficulties. There is no call for us to consider every one. But we ought at least to grant to Darwin all credit for so doing. Always he shows himself to be totally honest and truth-seeking. Never is he more eager to sell his own product than to discover – let the chips fall where they may – what actually is the case. In summarizing this chapter he says of the difficulties and objections: 'Many . . . are very grave, but I think that . . . light has been thrown on several facts, which on the theory of independent acts of creation are utterly obscure' (Darwin, C., 1859, p. 230).

(a) His attitude here is as usual a paradigm of scientific rationality, as well as of uncommon honesty and uncommon good faith. He does not ignore or attempt to hush up any apparently falsifying facts. These are all recognized and recorded among the difficulties and objections. But they are not mistaken as constituting decisive reasons to abandon what is by far the most promising theory available. Instead, he continues to develop, to defend, and to use that theory, but always in a suitably provisional, tentative and undogmatic way.

For, until and unless someone is able to think up a better theory it is entirely reasonable to hope that the apparent falsifications are not, after all, real and final; to hope, that is to say, either that the apparently falsifying facts are not truly facts, or else that, although they are, the theory can nevertheless eventually be so amended as to embrace and explain those formerly recalcitrant facts. Of these two possibilities the second, as we have just seen, was realized in the matter of difficulties in the area of variation; the first, as we soon shall be seeing, with respect to Lord Kelvin's objections that the earth is too young to have permitted a sufficiently protracted evolutionary past.

(i) Let us examine no more than two examples from Chapter VI. First, Darwin – who in the following year confessed in a letter to Asa Gray that he well remembered 'the time when the thought of the eye made me cold all over' (Darwin, F., II p. 296) – here gently rejects this objection: 'Although the belief that an organ so perfect as the eye could have been formed by natural selection, is more than enough to stagger anyone; yet in the case of any organ, if we know of a long series of gradations in complexity, each good for its possessor, then, under changing conditions of life, there is no logical impossibility in the acquirement of any conceivable degree of perfection through natural selection' (Darwin, C., 1859, p. 231).

(ii) Second, Darwin again displays the explanatory power of his theory by indicating a further dimension of falsifiability: 'Natural selection will produce nothing in one species for the exclusive good or injury of another; though it may well produce parts, organs, and excretions highly useful or even indispensable, or highly injurious to another species, but in all cases at the same time useful to the owner' (ibid., p. 232).

In view especially of the recent brouhaha raised by Richard Dawkins in *The Selfish Gene*, and by others urging the supposedly inescapable selfishness of the whole process of evolution, it becomes worth pointing out at once that Darwin draws too positive an inference. Natural selection does not positively produce anything. It only eliminates, or tends to eliminate, whatever is not competitive. For some variant characteristic not to be thus eliminated, it does not need to bestow upon its possessor any actual competitive advantage. It is both necessary and sufficient that it should not burden that possessor with any competitive disadvantage.

Darwin's mistake – a mistake the correction of which leaves us still with a satisfactorily falsifiable implication – is perhaps consequent upon his employment of the expressions 'natural selection' or 'survival of the fittest', rather than his own ultimately preferred alternative 'natural preservation' (Darwin, F., II p. 346). Certainly it cannot be either too early or too often emphasized that natural selection is no more a kind of (conscious and grounded) selection than Bombay duck is a species of duck. Nor, of course, are genes, or plants, or any organisms other than specimens of certain higher animals, ever engaged, whether selfishly or unselfishly, in any conscious or chosen pursuit of anything.

(b) Chapter VII, 'Instinct', is presented as a response to a major objection: '. . . so wonderful an instinct as that of the hive-bee making its cells will probably have occurred to many readers' (Darwin, C., 1859, p. 234). Observing that in fact instincts do 'vary slightly in a state of nature', Darwin can 'see no difficulty, under changing conditions of life, in natural selection accumulating slight modifications to any extent, in any useful direction'. The theory also gains positive support from the fact that similar species inhabiting widely separated regions – sorts of thrush, for instance, and sorts of wren – often possess uncannily similar instincts. Finally, Darwin treats himself to an admittedly not strictly scientific Parthian shot, which we shall need to recall in later chapters of the present book. To his mind, he confesses, 'it is far more satisfactory to look at such instincts as the young cuckoo ejecting its foster-brothers, – ants making slaves, – the larvae of ichneumonidae feeding within the live bodies of caterpillars, – not as specially endowed or created instincts, but as small consequences of one general law, leading to the advancement of all organic beings, namely, multiply, vary, let the strongest live and the weakest die' (ibid., p. 263).

(c) Chapter VIII, 'Hybridism', argues that the facts here seem not to be opposed to, 'but even rather to support the view, that there is no fundamental distinction between species and varieties' (ibid., p. 290). The next two chapters, 'On the Imperfection of the Geological Record' and 'On the Geological Succession of Organic Beings', form a pair. The imperfections of which Darwin spoke, thus, quietly begging the crucial question, have in the years between been progressively remedied; although, since only the most minute fraction of all once living creatures could have chanced to become fossilized, that record can never even begin to approach completeness. He himself stressed the gravity, both of the general lack of 'infinitely numerous transitional links between the many species which now exist or have existed', and in particular, 'the almost entire absence, as at present known, of fossiliferous formations beneath the Silurian strata' (ibid., p. 315).

(i) For our contemporaries the most remarkable thing about Chapter IX is that, whereas nowadays palaeontology and the geological record provide a main support for evolutionary theory, Darwin's concern was, rather, to show that they need not be regarded as opposing it (Romer). One of the most spectacular of 'missing link' discoveries occurred in 1861, in time for a mention in later editions of *The Origin of Species*. This was the fossil Archaeopteryx, intermediate between lizard dinosaurs and feathered true birds, a specimen of which turned up in an appropriate rock formation in Germany. More recently, members of the Leakey family have unearthed in Africa, as Darwin was later to forecast in *The Descent of Man*, bones belonging to various recent ancestors of homo sapiens. And, in our century again, we have increasing fossil evidence testifying unambiguously to the existence of living organisms in the vast and, so far as Darwin knew, almost empty pre-Cambrian period (Ruse, 1982, pp. 165ff.).

(ii) Chapter X shows how this theory, and no other, can explain the facts of succession. For instance: 'When a group has once wholly disappeared, it does not reappear; for the link of generation has been broken' (Darwin, C., 1859, p. 342). Again: 'We can understand how the spreading of the dominant forms of life, which are those that oftenest vary, will in the long run tend to people the world with allied, but modified, descendants; and these will generally succeed in taking the places of those groups of species which are their inferiors in the struggle for existence . . . The inhabitants of each successive period in the world's history have beaten their predecessors in the race for life, and are, in so far, higher in the scale of nature; and this may account for the vague yet ill-defined sentiment, felt by many palaeontologists, that organization has on the whole progressed. If it should hereafter be proved that ancient animals resemble to a certain extent the embryos of more recent animals of the same class, the fact will be intelligible' (ibid., pp. 342 and 343).

(d) Chapters XI and XII again form a pair, dealing with 'Geographical Distribution'. Here the impact of *The Voyage of the 'Beagle'* is clear and decisive: 'In considering the distribution of organic beings over the face of the globe, the first great fact which strikes us is, that neither the similarity nor the dissimilarity of the inhabitants of various regions can be accounted for by their climatal and other physical conditions' (ibid., p. 344).

(i) The point is that the present, to Darwin familiar, facts of distribution can only be accounted for if we postulate descent and evolution in the past: both periods when migrations were possible; and periods of isolation. Since no species has ever originated in two different places, even the most similar environments, if effectively isolated one from the other, have different fauna. If this is not so, 'why do we not find a single mammal common to Europe and Australia or South America?' (ibid., p. 349). The seeds of plants on the other hand, as Darwin and others had shown by experiment, can be carried over obstacles which could not but block mammalian migration.

(ii) These two chapters are, thanks to the *Beagle* experience, among the most overwhelming. The same could also be said about the penultimate Chapter XIII, 'Mutual Affinities of Organic Beings: Morphology: Embryology: Rudimentary Organs'. This deals with the sort of facts, some of which are familiar to everyone, which should make the thesis of individual and separate creations of all species wholly unbelievable; the sort of facts which had for centuries led taxonomists to continue to employ in their classifying – the strength of their commitment to that thesis notwithstanding – familial notions the appropriateness of which they thus felt bound to deny.

These facts are, in a word, the facts of homology: '. . . the members of the same class . . . resemble each other in the general plan of their organization. This resemblance is often expressed by the general term "unity of type"; or by saying that the several parts and organs in the different species of the class are homologous . . . What can be more curious than that the hand of a man, formed for grasping, that of a mole for digging, the leg of the horse, the paddle of the porpoise, and the wing of the bat, should all be constructed on the same pattern, and should include the same bones in the same relative positions?' (ibid., p. 415).

What indeed? And what could, what can, the so long and so confidently sought natural system of classification be if it is not the one in which positions and memberships are determined by the fundamental familial relationships of mating and parenting; those very relationships which systematists always in fact have allowed to determine that often very different males and females, parents and offspring, all belong to one and the same species? So, Darwin concludes, 'the several classes of facts which have been considered in this chapter, seem to me to proclaim so plainly, that the innumerable species, genera, and families of organic beings, with which this world is peopled, have all descended, each within its own class or group, from common parents, and have all been modified in the course of descent, that I should without hesitation adopt this view, even if it were unsupported by other facts or arguments' (ibid., p. 434).

(e) It remains only to write Chapter XIV, 'Recapitulation and Conclusion', and Darwin can rest his case. However, before proceeding to our own second chapter, we have to notice two further, substantial objections, to both of which decisive replies can now be given.

(i) First, in a formidable series of papers from 1862 onwards the physicist Sir William Thomson (later Lord Kelvin) argued that the geologists had to be wrong about the age of the earth and, hence, that there had not been enough time for species to originate by natural selection. Given the classical laws of heat-production and radiation, he calculated the rate at which the earth and the sun should be cooling through the loss of their own natural heat, and this calculation set upper limits to the ages of both the earth and the sun. The sun could not have existed for more than 500 million years, while the earth could not have supported life for more than a few million years.

Since the physical calculations could not be faulted it was clear that something had to go, or to come. Either evolution by natural selection would have to go, or there would have to be a revolution in physics, or Darwin's theory would have to be reinforced by the inclusion of some provision for making the occurrence of favourable variations much more frequent than Darwin himself believed that they were. He was never able to see his way round this particular obstacle, and wrote in his final revision of the *Origin*: 'With respect to the lapse of time not having been sufficient . . . this objection, as urged by Sir William Thomson, is one of the gravest as yet advanced. I can only say, firstly, that we do not know at what rate species change as measured by years, and secondly, that many philosophers are not as yet willing to admit that we know enough of the constitution of the universe and of the interior of our globe to speculate with safety on its past duration.'

In the event, Darwin has been totally vindicated in his altogether rational refusal to abandon a theory for which he had, as we have seen, an enormous mass of supporting evidence; and to which there was not, and never has been, any scientifically presentable alternative. On the one hand, the development of genetics has shown that the units of variation, taking the form of mutations, can be much bigger than Darwin was inclined to believe; and the whole evolutionary process hence much faster. (We should perhaps also mention punctuated equilibria: the suggestion of Stephen

Jay Gould and others that evolution by natural selection proceeds in relatively short, swift bursts interspersed with relatively long periods of relative equilibrium.) On the other hand, the hoped-for revolution in physics has in fact occurred. The discovery of radioactivity, more than a dozen years after Darwin's death, showed that he and his contemporaries did not 'know enough of the constitution of the universe and of the interior of our globe to speculate with safety on its past duration'. In 1901, Joly demonstrated that heat from the radioactive minerals within the earth could by itself slow down the planet's rate of cooling to such an extent that its probable life was vastly longer than the 24 million years estimated by Kelvin. By 1931, his estimate of the earth's age had been multiplied by one hundred, while nowadays the accepted date for the condensation is a good 5,000 million years back. His estimates of the age of the sun have been similarly shattered: indeed, the fusion of hydrogen atoms to form helium releases so much energy that the sun may not be cooling down appreciably at all. Darwin could not have dared to hope for more.

(ii) The second substantial objection to be noticed here was put, in a nutshell, by the geologist Adam Sedgwick, one of Darwin's most distinguished non-converts: '. . . as to your grand principle – *natural selection* – what is it but a secondary consequence of your supposed or known, primary facts?' Sedgwick's point was, as far as it went, quite correct. In so far as Darwin had demonstrated that evolution by natural selection must have occurred, and must still be continuing, he had not pointed to any examples of it visibly proceeding before our eyes. But several such examples are now available.

The most often cited is the phenomenon of industrial melanism, the appearance, particularly in Britain, of dark varieties of many species of moths. This phenomenon, first rigorously studied by J. B. S. Haldane in 1924, has more recently by H. B. D. Kettlewell been shown to be a most elegant example of the effect of a change in the environment on the genetical structure of populations. In areas subject to atmospheric pollution dark varieties are at an obvious competitive advantage as against light, and tend rapidly to replace them. (No doubt we shall soon be able, thanks to a combination of Clean Air Acts and progressive deindustrialization, to read equally persuasive papers about the reverse process.)

Another similar demonstration was reported by Dobzhansky in 1951. Darwin had noted that insects on islands are often distinctive in being wingless or in having only poorly functioning wings, and had deduced that, since flies with no or poor wings are less likely to be blown out to watery deaths at sea, selection must in such circumstances have favoured the less well equipped varieties. So Dobzhansky set up experimental cages of fruit-flies (Drosophila again), mixing flies with wings with flies without wings. Under normal conditions, as expected, flies with wings far out-reproduced flies without wings. But, after a current of air had been blown through the cages, the result was, as expected, reversed.

II

THE PHILOSOPHICAL IMPLICATIONS

The Origin of Species . . . With the one exception of Newton's *Principia* no single book of empirical science has ever been of more importance to philosophy than this work of Darwin.

Josiah Royce, *The Spirit of Modern Philosophy*, p. 286.

The Darwinian theory has no more to do with philosophy than has any other hypothesis of natural science.

Ludwig Wittgenstein, *Tractatus Logico-Philosophicus*, § 4.1122.

When the word 'philosophy' in each of these two apparently contradictory sentences is given the appropriate sense, then both express plain and entirely compatible truths, truths which are both in their different contexts important. In the first the relevant sense is wide and untechnical. It is in this original and most common understanding that biographers devote chapters to the philosophy of their subjects, professional associations invite leading figures to address ceremonial occasions on their personal philosophy of whatever it may be, and editors of serious general journals commission contributions to symposia on the philosophical implications of new theoretical developments. In the second sentence quoted the relevant sense is narrow and technical. In this specialist sense we could with equal truth say: that, while most of *The Laws* and much of *The Republic* is not philosophy, Plato's equally characteristic but less read *Theaetetus* is almost nothing but; and that, of the comparatively few pages of such philosophy in Hume's second *Inquiry*, most are of intent relegated to the Appendices (Flew, 1979, ch. I).

The two understandings of philosophy thus illustrated should both be acceptable. Although they are very different, the second often has implications for the first. Indeed, in the present chapter they will be seen to be very much mixed up together and sometimes not distinguishable one from the other at all. The first section, for instance, is concerned with what is very much the sort of questions asked by those who are by profession philosophers in the narrower sense. Yet certain popular but wrong answers to some of these questions must tend to undermine the standing of Darwinism as science, and hence to diminish or to destroy, in the untechnical sense of the word 'philosophical', its philosophical significance. Indeed, that section is going to begin with a consideration of a recent onslaught upon the theory of evolution by natural selection, an onslaught based primarily on philosophical misconceptions of the narrower kind.

1. THE LOGICAL STRUCTURE AND FUNCTIONS OF DARWIN'S THEORY

In 1980 and 1981 both *Nature* and *New Scientist* – the two journals which circulate most widely within the British scientific community – published editorial comments on and a deal of correspondence about a newly organized exhibition in the great national Natural History Museum in South Kensington, London. One aspect of the matter to which the editors and several correspondents rightly took exception was that the staff members responsible – all, by the way, civil servants – had presented the exhibits within the framework of their own pet theory, called cladism or cladistics, without warning the public that this was what they were doing. We shall return to this somewhat elusive new theory later. The objection, of course, was not to their holding and arguing for what is in fact very much a minority view, but rather to their abuse of the resources of a state institution to try to put that view across to all the innocent and predominantly youthful laypersons who throng these public galleries, as if it were already part of the established consensus among all those best qualified to judge. But our immediate concern is with some of the assertions in the equally anonymous supporting materials. (The quotations which follow are all borrowed from an article by Barry Cox in *Nature* for 4 June 1981, p. 373. The offending material has since, apparently and none too soon, been withdrawn.)

(a) 'The Survival of the Fittest', they said,

is an empty phrase; it is a play on words. For this reason, many critics feel that not only is the idea of evolution unscientific, but the idea of natural selection also. There's no point in asking whether or not we should believe in the idea of natural selection, because it is the inevitable logical consequence of a set of premises . . . The idea of evolution by natural selection is a matter of logic, not science, and it follows that the concept of evolution by natural selection is not, strictly speaking, scientific.

(i) The pretended sophistication of both these statements is, in fact, to be plain, just wrong. First, since the criterion of fitness to survive is in this case actual survival, the assertion that the fittest do in fact survive is certainly not to be construed, as we all know that too often it has been construed, as a guarantee that in the natural order all is for the best. But from this it does not at all follow that that assertion is tautological. It is not. For clearly it denies that the survivors survive at random, while maintaining that they have some kind of competitive edge over the non-survivors.

The trouble here, I suggest, is that people have rushed forward to announce the detection of tautology without taking time to specify precisely what proposition it is which is supposed to be (merely) tautological. Such impetuous carelessness used to be a very common fault among professional philosophers, paid to know better. The present particular variety of this fallacy species still has, to continue an appropriately biological form of speech, an extremely wide distribution: '. . . the number of critics of this kind is legion' (Ruse, 1973, p. 30).

(ii) Second, it is not merely careless but plumb preposterous to tell us that something which 'is the inevitable logical consequence of a set of premises' thereby becomes 'a matter of logic, not science'. To appreciate the point of this antithesis between logic and science we need to recall another of the gnomic apothegms of Ludwig Wittgenstein, as rendered in the Authorized Version of his *Tractatus Logico-Philosophicus*: 'But all propositions of logic say the same thing. That is, nothing' (§ 5.43, p. 121). Such analytic propositions are, in Humean terms, those which state or purport to state only 'the relations of ideas'; whereas science is understood to include nothing but propositions of the other basic sort – those stating or purporting to state 'matters of fact and real existence' (Hume, 1748, IV i p. 25). The former sort are, if true, necessarily and tautologically true and, if false, self-contradictory; whereas the contradictories of propositions of the latter sort are never self-contradictory, even when these propositions themselves happen to be true.

To illustrate and to fix this fundamental distinction, contemplate an exchange in *Hamlet* (1.5). Questioned about the ghost the Prince replies:

> There's ne'er a villain dwelling in all Denmark
> But he's an arrant knave.

Horatio, a shade nettled, responds:

> There needs no ghost, my lord, come from the grave
> To tell us this.

There needs no ghost, because the proposition which Hamlet has uttered is tautological. It is thus necessarily true, expressing no more than the formal relations of ideas. So it can be known to be true, albeit empty, without research, by anyone who is master of the modest range of notions employed therein. Contrast the different script to be provided by a sociologist Shakespeare – a sociologist without, I fear, much respect for metre, or anything else:

> There's ne'er a villain dwelling in all Denmark
> But he or she is a guiltless victim of whoever
> are my own favourite enemies.

To know that any such claim about 'matters of fact and real existence' is true or, more likely, false someone would have to do or to have done some actual face-to-face fieldwork. And, even if it were true, it must remain at least conceivable that it might have been false.

Once these explanations and distinctions are mastered it becomes easy to trace the source of the trouble. The unstated and hence unquestioned false assumption is, that all valid demonstrative arguments must proceed from tautologically true premises to tautologically true conclusions. There are two errors here, errors peculiarly discreditable in publicizers and interpreters of scientific work. One is to assume that both the premises and the conclusions of an argument have to be true if that argument is to be demonstratively valid. The other is to assume that such demonstrative arguments can only proceed from tautologically true premises to tautologically true conclusions. What makes these errors in professional publicizers of scientific work so peculiarly inept is that the essentially hypothetico-deductive procedures of science would become altogether impossible if either assumption were true. For these procedures involve, indeed they are, the deducing of consequences which would, or will, obtain if the hypothesis under discussion were to be, or is, true; with always the very live possibility of discovering that the consequence actually does not obtain, and of thereby discrediting the hypothesis from which it was thus validly derived. Plainly, all this would be impossible if there could not be valid deductions from premises and to conclusions which purported to state, but which were in the event discovered not truly to state, 'matters of fact and real existence'.

(b) Before proceeding to expose further errors propagated by the cladist cadre at the Natural History Museum, we need to look again and more closely at the deductive core of Darwinism, in order the better to appreciate in this present context the character both of the 'set of premises' and of the 'inevitable logical consequences' derived therefrom. One modern interpreter maintains a thesis which might seem to warrant a damping conclusion:

> The old arguments for evolution were only based on circumstantial evidence . . . But the core of Darwin's argument was of a different kind. It did not make it more probable – it made it a certainty. Given his facts his conclusion *must* follow: like a proposition in geometry. You do not show that any two sides of a triangle are very *probably* greater than the third. You show that they *must* be so. Darwin's argument was a *de*ductive one – whereas an argument based on circumstantial evidence is *in*ductive. (Pantin, p. 137)

Julian Huxley once took a similar hint from the writings both of Darwin himself and of his own grandfather T. H. Huxley. Neo-Darwinism, the younger Huxley then suggested, can 'be stated in the form of two general evolutionary equations. The first is that reproduction plus mutation produces natural selection; and the second that natural selection plus time produces the various degrees of biological improvement that we find in nature' (Huxley, J., 1953, p. 38). The idea is excellent, but the execution here is curiously slapdash. For the first equation, as Huxley gives it (R + M→NS), is not valid. Reproduction plus mutation would not necessarily lead to natural selection. It is essential to bring in the struggle for existence; and that in turn has to be derived from the sum of the geometrical ratio of increase plus the limited resources for living – limited *lebensraum* as Hitler's Germans might have called it. So, to represent the core of the deductive argument, we need something more like: GRI + LR→SE; SE+ M→NS: and, not on all fours with the first two, NS + T→BI. To represent Darwinism rather than Neo-Darwinism V (hereditable variation) must be substituted for M (mutation).

Of course, in order to make this deductive core of Darwin's argument, and the equations employed to represent it, ideally rigorous we would have to construct for all the crucial terms definitions to include explicitly every necessary assumption. There are in fact several, many of which when uncovered may seem too obvious to have been worth stating. Take, for instance, one to which Darwin himself refers: 'I think it would be a most extraordinary fact if no variation had ever occurred useful to each being's welfare' (Darwin, 1859, p. 170). It would indeed. Nevertheless, if we are to have a rigorously demonstrative argument, all the required premises have to be in one way or another made explicit. In fact, Darwin at once offers a very powerful reason for believing that the missing premise is true. For he adds after the phrase just quoted: 'in the same manner as so many variations have occurred useful to man'. In this case, bringing out the assumption has a value other than that of rigorization for its own sake. For it should suggest, what has in fact proved to be the case, that one of the main effects of natural selection is to eliminate unfavourable variations. It not only helps to generate biological improvement. It is essential to prevent biological degeneration.

However, it is not necessary for us to attempt to construct an ideally rigorous formalization, satisfying the austere canons of contemporary symbolic logic. For, even if we could succeed in this task, and even if the result was in some way an improvement on anything provided in *The Origin of Species*, still it would necessarily be, as an improvement, a misrepresentation of what was actually said there (Oldroyd, pp. 117ff.).

(i) Although that argument proceeds deductively or, if you like, *a priori*, it most certainly does not follow that it is – to coin some charges! – 'a matter of logic, not science', a farrago of empirically 'empty phrases', nothing but 'a play on words'. For, since its premises all state 'matters of fact and real existence', its conclusions are of the same kind.

That living organisms all tend to reproduce themselves at a geometrical ratio of increase; that the resources they need to sustain life are limited; and that while each usually reproduces after its kind, sometimes there are variations which in their turn usually reproduce after their kind: all these propositions are none the less contingent and empirical for being manifestly and incontestably true. That there is a struggle for existence; and that through this struggle for existence natural selection occurs: both these propositions equally are none the less contingent and empirical for the fact that it follows, necessarily as a matter of logic *a priori*, that wherever the first three hold the second two must hold also. (A proposition is contingent if its contradictory is not self-contradictory, and empirical if it makes some substantial claim about actual or possible experience. For present and perhaps for all purposes the class of the contingent and the empirical may be identified with Hume's 'matters of fact and real existence'.)

(ii) The premises are matters of obvious fact. The deductive steps are short and simple. The conclusions are enormously important. Yet these conclusions, and that they were implied by these premises, were before Darwin very far indeed from being obvious to able men already sufficiently familiar with those premises. This should give us a greater respect for the power of simple logic working on the obvious. Of course, he did not just have to make some short deductive moves from a few very wide-ranging empirical premises already provided as such. He had first to recognize that these propositions did constitute essential premises; and then, after making the deductions from them, to appreciate that these premises and these conclusions contained and linked together concepts crucial for understanding the problem of the origin of species. The premises, the concepts, the deductions, the conclusions, all are simple. To bring them together and to see the importance of the theoretical scheme so constructed was a simple matter too. Yet this simplicity is the simplicity of genius.

(iii) The conclusions of the deductive argument are proved beyond dispute. For, though the premises are as empirical generalizations in principle open to revision, in fact they cannot reasonably be questioned. It is therefore all the more important to appreciate what it does not prove; and, hence, what, at least as far as this argument is concerned, is left open to be settled by further inquiry.

It certainly does not prove that all 'the various degrees of biological improvement that we find in nature' can be accounted for in these terms (Huxley, J., 1953, p. 38). At most it proves only that some 'biological improvement' must occur. It would be wholly compatible with this deductive core to maintain that all or most new species are in fact specially created; notwithstanding that any such arbitrary and anti-scientific postulations would be entirely out of harmony with Darwin's own thoroughly naturalistic spirit, and his Lyellian insistence on continuity of development: '. . . species are produced and exterminated by slowly acting and still existing causes, and not by miraculous acts of creation and by catastrophes' (Darwin, C., 1859, p. 57). Here, however, we should perhaps remind ourselves that in *The Origin of Species* he always concedes a special start for the first life: '. . . life, with its several powers, having been breathed into a few forms or into one' (ibid., pp. 459–60); although 'I should infer from analogy that probably all the organic beings which have ever lived on this earth have descended from some one primordial form, into which life was first breathed' (ibid., p. 455; in later editions references to the creator were added).

Darwin therefore needed and, as we saw in Chapter I, provided considerations of a different kind to support his own far wider and more revolutionary conclusion, a conclusion which cannot be enforced by the deductive core argument alone: '. . . species are not immutable; but that those belonging to what are called the same genera are lineal descendants of some other and generally extinct species, in the same manner as the acknowledged varieties of any one species are the descendants of that species . . . I am convinced that Natural Selection has been the main but not exclusive means of modification' (ibid., p. 69). And, although he built up a mighty case for this sweeping conclusion, that case is one which could not in principle be complete unless the whole science of evolutionary biology were complete. Here the field for inquiry is open and without limit.

(c) Another exceedingly common and endarkening error was both made and propagated by these officials at the Natural History Museum; officials, we may wryly note, charged with a duty of public enlightenment. It is that of assuming that no theory and no proposition describable as theoretical, can be either known to be, or even with overwhelmingly good reason believed to be, true. (I am reminded of the first-year student at the University of Keele who, asked what evidence there is for the theory of evolution by natural selection, replied: 'There isn't any evidence. It's just a scientific theory.' That particular student, I am happy to say, later redeemed himself by achieving first-class honours in mathematics and physics.)

(i) This false and obscurantist assumption can be refuted perhaps most elegantly and most agreeably by reflecting how regularly the theories of Sherlock Holmes were shown to have embraced the simple or sometimes complicated truth about who did it, and how. But we need also to discover, and to seal or sterilize, the sources of error. In one of the exhibition brochures, which typically refers to cladistic theory as if this was an element in the consensus of the experts, we read: 'Biologists try to reconstruct the course of evolution from the characteristics of living animals and plants from fossils, which give a time scale to the story. If the theory of evolution is true . . .'

This drew a suitably magisterial rebuke from the editor of *Nature*:

> Can it be that the managers of the museum which is the nearest thing to a citadel of Darwinism have lost their nerve, not to mention their good sense? . . . Nobody disputes that, in the public presentation of science, it is proper whenever appropriate to say that disputed matters are in doubt. But is the theory of evolution still an open question among serious biologists? And if not, what purpose except general confusion can be served by these weasel words? (26 February 1981: vol. 289, p. 735)

This editorial was later shown to have been the more necessary when no less a person than the then President of the Royal Society, himself another grandson of Thomas Henry Huxley, intervened to reproach *Nature* for having opened its columns to the previous correspondence on what is and is not going on in the museum, and why.

The associated filmclips contain phrases indicating some of the reasoning behind the conclusion that no theory can be known to be true: 'If we accept that evolution *has* taken place, though obviously we must keep an open mind on it . . .'; 'We can't prove that the idea is true, only that it has not yet been proved false'; and 'It may one day be replaced by a better theory, but until then . . .' Later, *Nature* published a letter from a group of 'scientists working at the British Museum (Natural History)', presumably (some of) those responsible both for the exhibition and the filmclips. This letter attempted to rebut the previously quoted rebuke from the editor:

> How is it that a journal such as yours that is devoted to science and its practice can advocate that theory be presented as fact? This is the stuff of prejudice, not science, and as scientists our basic concern is to keep an open mind on the unknowable. Surely it should not be otherwise? You suggest that most of us would rather lose our right hands than begin a sentence with the phrase 'If the theory of evolution is true . . .' Are we to take it that evolution is a fact, proven to the limits of scientific rigour? If that is the inference then we must disagree most strongly. We have no absolute proof of the theory of evolution. What we do have is overwhelming circumstantial evidence in favour of it and as yet no better alternative. But the theory of evolution would be abandoned

tomorrow if a better theory appeared. (12 March 1981: vol. 290, p. 82)

(ii) A refutation of all this at the level of common sense has already been provided in the previous sub-section, both by the editor of *Nature* and in the initial reference to the successful cases of Sherlock Holmes. But more is needed. In the first place, we must insist again upon a distinction drawn already in Chapter I. It is wrong to identify either the Darwinism of *The Origin of Species* or Neo-Darwinism with biological evolution without prefix or suffix. That to which any account of the evolution of species is necessarily opposed is any doctrine of their immutability; combined, presumably, with the claim that they were, whether simultaneously or successively, specially created by *ad hoc* supernatural agency. Darwinism and Neo-Darwinism, on the other hand, are accounts of how evolution has in fact occurred, and is continuing to occur, and here, presumably, the conceivable alternatives would include revivals or developments of what was offered in Lamarck's *Philosophie zoölogique* or in Chambers's *Vestiges*.

It is hard to believe that these British civil servants really are concerned to urge everyone 'to keep an open mind' about evolution, and to be ready to abandon the whole notion tomorrow in favour – would it be? – of the self-styled 'scientific creationism' of the fundamentalist ultras in the USA. (Biblical fundamentalists are those who believe in 'the precious book from cover to cover', very literally interpreted, and the description derives from a statement of 'fundamental principles' adopted in 1895 by the Niagara Bible Conference.) But if it is the Neo-Darwinist account of how such evolution has occurred and continues to occur about which, despite the admittedly 'overwhelming circumstantial evidence in favour of it', they want us to keep our minds open, then we must press the question: 'What is the possible fresh evidence for which you are asking us to wait?'

It would be a salutary imaginative exercise to try to describe the possible discoveries which would show that, after all, there is no such process as natural selection and that genetics is nothing but a pseudo-science. For nothing less than upset discoveries on this unprecedented and cataclysmic scale would suffice to warrant the alternative apparently entertained by the Natural History Museum. Certainly it would not be enough to point out, what equally certainly ought to be pointed out, that there may and very likely will have to be further additions and amendments, as extensive as those involved in the replacement of Darwinism by Neo-Darwinism. It is even possible to suggest where this next major supplementation will occur.

For evolutionary biologists have long been aware of very big differences in the apparent pace of change, most notably in a tumultuous few million years at the beginning of the Cambrian epoch. More particularly, we read in a recent survey: 'Even within the staid horse family, which seems as a whole to be progressing rather steadily through the Tertiary, close examination shows the rates varying considerably. More broadly, evolution commonly seems to proceed in spurts and pauses in an apparently erratic way' (Simpson, p. 103). Such observations have led some, perhaps especially those with Marxist affiliations, to deny 'the canon of "*Natura non facit saltum*"' and to insist that here too we can find examples of what the late Chairman Mao called 'a Great Leap Forward'. But what is at stake is not the rejection of Neo-Darwinism, only a possible amendment. Thus, the most brilliant of the innovators, Stephen Jay Gould – Professor of Biology, Geology and the History of Science at Harvard – was, while mentioning in an aside that he had 'learned his Marxism literally at his daddy's knee', careful to call his article 'Punctuated Equilibria: the Tempo and Mode of Evolution Reconsidered'. There is no doubt but that Gould would acknowledge a responsibility sooner or later to indicate some natural mechanism capable of producing these apparently erratic spurts and pauses – the biological analogue, as it were, of the irreversible seizure of absolute social power by the élite of a Marxist-Leninist party.

Our for ever open-minded correspondents from the Natural History Museum are not, however, really waiting for some crucial fresh evidence which even they would allow to be decisive. They are, on the contrary, adopting a universal philosophical position; and one which commits them to denying that any theory, and perhaps any proposition at all, can be known to be true: 'We can't prove that the idea is true, only that it has not yet been proved false'; and so on. And this philosophical position, like their earlier dismissal of all evolutionary theory as tautological, derives from Popper. The difference is that, whereas Popper has now repudiated that dismissal (Popper, 1978, p. 344), he has not yet found a way to escape the most paradoxical implications of *The Logic of Scientific Discovery*.

In that great book Popper wrote:

> The old scientific idea of *epistēmē* – of absolutely certain, demonstrable knowledge – has proved to be an idol . . . every scientific statement must remain *tentative forever* . . . The wrong view of science betrays itself in the craving to be right; for it is not his *possession* of knowledge of irrefutable truth, that makes the man of science, but his persistent and recklessly critical *quest* for truth. (Popper, 1959, pp. 280–1)

The clear implication seems to be that there is no such thing as scientific or any other kind of knowledge. And this he says in as many words later in the same book. For he proceeds to maintain 'that we must not look upon science as a "body of knowledge", but rather as a system of hypotheses . . . a system of guesses . . . of which we are never justified in saying that we know that they are "true" or "more or less certain" or even "probable"' (ibid., p. 317).

But elsewhere, even in that same book, this official doctrine seems to be forgotten: 'It is not truisms which science unveils. Rather, it is part of the greatness and beauty of science that we can learn . . . that the world is utterly different from what we ever imagined – until our imagination was fired by the refutation of our earlier theories (ibid., p. 431). Later, in *Conjectures and Refutations*, Popper commends (not instrumentalist but) realist theories, which 'have immeasurably extended the realm of the known. They have added to the facts of our everyday world the invisible air, the antipodes, the circulation of the blood, the worlds of the telescope and the microscope, of electricity, and of tracer atoms showing us in detail the movements of matter within living bodies' (Popper, 1963, p. 102). A mere three pages further on he is, if anything, even more emphatic: 'I cannot accept an argument that leads to the rejection of the claim of science to have discovered the rotation of the earth, or atomic nuclei, or cosmic radiation, or the "radio stars"' (ibid., p. 105). Finally, with *Objective Knowledge* we arrive at a book whose very title contradicts the official, Xenophanean doctrine that 'all is but a woven web of guesses' (ibid., p. 153). Yet nowhere can we find any attempt to introduce limitations on the scope of the grotesquely paradoxical official teaching of *The Logic of Scientific Discovery*.

This is not the place to try to draw the teeth of this fundamental Popperian paradox, something which I have in any case attempted elsewhere (Flew, 1982a, Part II; and compare Flew, 1961, ch. IV). It will be sufficient to make just two brief remarks. First, it is a perennial yet ruinous mistake to hold that for anyone to know, truly to know, it must be either inconceivable that things might have been other than they believe them to be, or inconceivable that they themselves might be in error in believing what they do believe, or both. If this is what Popper means by *epistēmē*, then that is indeed a philosopher's will-of-the-wisp. For the truth is that to know it is sufficient to be disposed to assert what is in fact true, and to possess full warrant for so doing. It is, of course, in this altogether sensible and everyday reading of the expression 'to know' that Popper makes his robust rationalist claims to know so much which, on his official view, neither is nor could be known. Second, even if anyone is so stubborn as consistently to maintain that official position, still that would commit them to a general and systematic devaluing of all not yet falsified theories, rather than to picking on Neo-Darwinism in particular.

(d) The three previous sub-sections have all dealt with misconceptions which have, it has to be confessed, been promoted partly or primarily by philosophers. The present final sub-section of Section 1 will indicate two or three ways in which sound philosophy may increase our understanding.

(i) As Darwin himself saw clearly, the acceptance of any such account of the origin of biological species requires or permits changes in certain categorial concepts. The Darwinist is bound and entitled to abandon assumptions implicit in the previous usage of the words 'genus', 'species' and 'variety' and all their close semantic associates: 'When the views advanced by me in this volume . . . are generally admitted . . . there will be a considerable revolution in natural history. Systematists will be able to pursue their labours as at present; but they will not be incessantly haunted by the shadowy doubt whether this or that form be in essence a species' (Darwin, C., 1859, p. 455). Which, he continues with feeling: 'I feel sure, and I speak after experience, will be no slight relief.'

The crux is that he is by his theory committed to rejecting the notion that every organism is a charter member of some natural kind, and has as such its pre-appointed proper pigeon-hole, which it is the business of the systematist to discover. In the light of that theory, he goes on, '. . . I look at the term "species" as one arbitrarily given for the sake of convenience to a set of individuals closely resembling each other, and that it does not essentially differ from the term "variety" which is given to less distinct and more fluctuating forms. The term "variety" again, in comparison with mere individual differences, is also applied arbitrarily, and for mere convenience's sake' (ibid., p. 108; inverted commas supplied). 'Hereafter we shall be compelled to acknowledge that the only distinction between species and well-marked varieties is, that the latter are known, or believed, to be connected at the present day by intermediate gradations, whereas species were formerly thus connected . . . In short, we shall have to treat species in the same manner as those naturalists treat genera, who admit that genera are merely artificial combinations made for convenience. This may not be a cheering prospect; but we shall at least be freed from the vain search for the undiscovered and undiscoverable essence of . . . species' (ibid., pp. 455 and 456).

The assumptions which Darwin's theory thus commits him to challenge are rather more particular cases of those completely general prejudices about language and classification which seem first to have been extensively challenged in Book III of John Locke's *Essay Concerning Human Understanding*. Although this *Essay*, first published in 1690, was enormously influential there appears to be no evidence that Darwin ever read any of Locke's more philosophical writings. We do however know: both that his grandfather Erasmus both used and recommended the use of commonplace books organized on principles derived from Locke's *Letters on Study* (Gruber, pp. 23, 263 and 317); and that his own unusually catholic reading included many authors influenced by Locke, although not necessarily on presently relevant issues.

The assumptions challenged, first by Locke and then by Darwin are: that all natural objects belong to certain natural kinds, in virtue of their 'essential natures'; that there are no marginal cases falling outside and between these sharply delimited collections of individuals; that there must always be straight yes or no answers to the question 'Is this individual a so and so or not?'; and that men have only to uncover, and, as it were, write the labels for, the classes to which God or nature has antecedently allocated every such individual. This conceptual aspect of Darwin's work may be compared with Einstein's analysis in his relativity theories of the ideas of motion and of simultaneity. In both cases the scientist has been to a greater or lesser extent anticipated by a philosopher. But, whereas Darwin seems never to have read the relevant parts of Locke, Einstein had read Mach; and acknowledged indebtedness immediately to him and ultimately to Hume.

Before we allow ourselves to become puffed up with the proud parochial superiority of hindsight, it is well to pause for a moment to notice the enormous initial plausibility of what we might label the Genesis view of the nature and presuppositions of biological classification. For a strong inclination to this view is by no means peculiar to way-out fundamentalist Bible bigots. On the contrary: it seems to be supported by all the most immediate everyday experience. Almost every animal or plant which forces itself on the attention of the biological layperson belongs to some commonly named kind very obviously and very widely different from every other ordinary familiar kind. Who would ever mistake a cat for a dog, a cow for a horse, or – come to think of it – a hawk for a handsaw? The same, of course, applies to specimens of all the kinds actually listed in Genesis. This first impression of the absolute distinctiveness of natural kinds is reinforced by the recognition that couples from all these kinds normally reproduce, as Genesis has it, after their own kind. And, furthermore, those hybrid creatures which are frequently found – mules, for instance – are sterile: hybridization thus appears to be unnatural; with sterility as nature's penalty for this unnatural offence.

(ii) Darwin's theory as expounded in *The Origin of Species* employs no unexplained technical terms, postulates no hypothetical entities, and involves no mathematics. It therefore constitutes an ideal textbook example for showing how a good theory – and, in this case surely, one which is also in large part true – can both explain what was previously known yet puzzling, and stimulate and direct inquiries fruitful of fresh discoveries. The explanations consist in showing how, granted that the mechanisms were and are as Darwin argued that they were and must be, then, the phenomena to be explained are just what we should have expected. Thus in his final chapter, 'Recapitulation and Conclusion', he reviews some of the sorts of properly puzzling facts which he has earlier examined in greater detail: 'We can plainly see,' he writes, 'why nature is prodigal in variety, though niggard in innovation. But why this should be . . . if each species has been independently created, no man can explain' (Darwin, 1859, p. 445). Again, 'Looking to geographical distribution, we admit that there has been during the long course of ages much migration . . . then we can understand, on the theory of descent with modification, most of the great leading facts in Distribution . . . we can understand, by the aid of the Glacial period, the identity of some few plants, and the close alliance of many others, on the most distant mountains, under the most different climates; and likewise the close alliance of some of the inhabitants of the sea in the northern and southern temperate zones, though separated by the whole intertropical ocean' (ibid., p. 449). And so it goes on.

Darwin himself recommended his theory on a second count, too. Not only does it offer true explanations, but it also promises to be fertile in discoveries of further truth. (How absurd it is, by the way, to say, as today it is so often said, that scientists are concerned not for truth but only for heuristic fertility. For of what are fertile theories thought to be fertile if it is not discoveries of what, for better or for worse, actually is the case, is true?) The whole great promise is summed up in an eloquent passage in that same final chapter: 'A grand and almost untrodden field of enquiry will be opened, on the causes and laws of variation . . . A new variety raised by man will be a more important and interesting subject of study than one more species added to the infinitude of already recorded species. Our classifications will come to be, as far as they be so made, genealogies . . . we have to discover and trace the many diverging lines of descent in our natural genealogies . . . Rudimentary organs will speak infallibly with respect to the nature of long lost structures . . . Embryology will reveal to us the structure, in some degree obscured, of the prototypes of each great class' (ibid., pp. 456–7).

Those references to genealogies should remind us of a sentence quoted already in Chapter I: 'The terms used by naturalists of affinity, relationship, community of type, paternity, . . . will cease to be metaphorical, and will have a plain signification' (ibid., p. 456). As usual Darwin appreciates exactly what he is doing, and why it is important. He is – to introduce a handy modern term – deploying an old model. Scientists habitually speak of trying to understand some range of phenomena in terms of this or that model, in terms, that is to say, of this or that analogy: light, for instance, is thought of as being or being like something which travels; while homologies lead biologists to talk of resembling species as members of a kind of family. (Homologies are resemblances of the sort to which Darwin was referring when he asked: 'What, for instance, is more wonderful than that the hand to clasp, the foot or hoof to walk, the bat's wing to fly, the porpoise's fin to swim should all be built on the same plan? and that the bones in their position and number that they can all be classed and called by the same names?') How far a model is said to be deployed is a matter of how far the analogy, between that model and the phenomena to which it is applied, is believed to hold.

The acceptance of Darwin's theory involves the deployment of a model which had been first applied a very long time before, but which had then remained curiously boxed up and impotent. This was, of course, the model of the family: with which were associated the methods of representation employed to express both familial relationships – the family tree – and one system of classification – the tree of Porphyry. (The Neo-Platonist Porphyry discussed and illustrated this in his influential Introduction to Aristotle's *Categories*, translated into Latin by Boethius.) The most remarkable fact about the familial model precisely is the enormous time-lag between its first introduction and its Darwinian development. The various terms appropriate to this model were, apparently, introduced because naturalists noticed analogies which made the idea of family relationship seem apt as a metaphor. But the suggestion that the metaphor might be considerably more than a mere metaphor, that the model could be deployed, seems almost always to have been blocked by the strong resistance of the accepted doctrine of the fixity of species. And this resistance must in its turn have been strengthened both by false conventional wisdom about the nature and presuppositions of biological classification, and by those inescapably familiar everyday realities which make that Genesis view so plausible.

Finally, before moving on to Section 2, it is worth returning for a last encounter with the anti-Darwinian cell at the Natural History Museum. Part of the fall-out of the controversy stirred by their new display was a BBC programme asking, 'Are the reports of Darwin's death exaggerated?', a programme the script of which was afterwards printed in *The Listener* for 8 October 1981 (vol. 105, no. 2730, pp. 390–2). The 'transformed cladism' expounded there by Colin Patterson, a senior member of the scientific staff of that museum, is, in the most literal sense, reactionary. For the transformed cladist is a born-again Aristotelian, who insists on classifying organisms without reference to their supposed evolutionary genealogy.

A good classification, we are told, looks at many characteristics, and it is important to know how much weight to give to each one, which are important and which aren't. How true, how true! Fortunately, cladistics is there to help us: 'Cladistics', we are assured, 'is a method of doing that. It makes the practice of taxonomy exacting because it forces the scientist to be objective and to sort out important from unimportant features of plants and animals.'

Unfortunately, no one offers to specify the interests by reference to which such importance or unimportance is supposed to be thus objectively determined. Yet to ask for an objective determination of what, without reference to any interests, is absolutely important would be no more sensible than to seek to know whether something is in motion absolutely, and not relative to anything else at all.

We can perhaps find a clue to what the unexplicated guiding interests of the cladists are in references to 'natural philosophers going back to Aristotle – most notably Linnaeus – who classified perfectly happily without any inkling of descent'. Their theory was of real essences or natural kinds: the real essence being, as it were, the blueprint used by God or nature in constructing all genuine specimens of that particular natural kind. Given that theory, then the importance of any characteristic is determined by its reliability and utility as an indicator of natural kind membership. So we have to conclude that even where Darwin is dead, Aristotle is very much alive.

2. THE CHALLENGE TO RELIGIOUS ASSUMPTIONS

Let us examine this challenge in four stages. In the first we shall consider how Darwin's scientific work and personal experience affected his views about religion. In the remaining three we shall inquire what evolutionary biology does and does not imply with regard to creation, design, and the uniqueness of man.

(a) We have already seen that Darwin boarded *Beagle* with a straightforward, literalistic belief in all the then teachings of the Church of England. We even know that in the first months of the voyage he was sometimes teased by some of his shipmates for the literalism of this faith. But this simple-minded Bible fundamentalism was progressively undermined: in part by his enthusiastic study of Lyell's *Principles of Geology*; in part by his own geological and biological observations; and in part, no doubt, by reaction against the implacable hard-line orthodoxy of Captain FitzRoy, whose cabin he shared. But this progressive undermining was so far only an undermining of the simple literalism. Halfway through the voyage he wrote in a letter: 'I often conjecture what will become of me; my wishes certainly would make me a country clergyman.' And he even co-authored with FitzRoy an appeal for support for Christian missionary work in the Pacific – 'The Moral State of Tahiti'.

The radical questioning began only after his return to England, and after the starting of the first notebook on transmutation in July 1837. Two of these transmutation notebooks, M and N, were written in, respectively, 1838 and 1839. Years later, in 1856, Darwin reread them, and declared that they were 'full of metaphysics on morals'. They have recently been published in an exhaustive and excellent critical edition. They make it quite clear that he was now moving beyond a modest rejection of the stories in Genesis 1 and 2, literally interpreted, and on to a sort of materialism. Although the immutability of species was, and for many years was to remain, the accepted orthodoxy, there had, as we have seen, been outspoken evolutionists before – Erasmus Darwin, for one. But materialism, with its insistence that stuff is prior to personality and consciousness, was far more fundamentally disturbing. It suggests that man is as much an element in nature as any of the brutes, and neither has nor can have any incorporeal and immortal destiny beyond the grave.

In a marginal note in his copy of John Abercrombie's just published *Inquiries concerning the Intellectual Powers* Darwin defined his term: 'By materialism I mean merely the intimate connection of kind of thought with form of brain' (Gruber, p. 201). The best short statement is not in either the M or the N notebook but in C, dating to May 1838: 'Love of deity effect of organization, oh you materialist! . . . Why is thought being a secretion in the brain more wonderful than gravity a property of matter? It is our arrogance . . . our admiration of ourselves' (ibid., p. 97).

The other main sources, indeed the only sources for the whole later period, are the *Autobiography* and the correspondence. Of correspondence there is plenty since, as has been noticed already, all concerned seem to have obeyed the Chinese maxim: 'Never kill the draught ox, nor throw away written paper.' The *Autobiography*, written in 1876, was first published in 1887 as Chapter II of the first volume of *The Life and Letters of Charles Darwin*, edited by his son Francis. But this version, mainly in order to spare the feelings of Charles Darwin's beloved and utterly devoted wife Emma, suppresses the most forceful passages about religion. These were only restored, and the text published in full, seventy-six years after his death. This is, of course, the definitive text, edited by a granddaughter Nora Barlow (née Darwin). The whole document is exceedingly good reading, not by any means only for these formerly excised passages.

Darwin's first reasons for abandoning belief in Christianity as a system of religious revelation are neither novel to him nor especially interesting; and 'The rate was so slow that I felt no distress' (Barlow, p. 87). It began with incredulity about the contents of the first chapters of Genesis, and continued 'By further reflecting that the clearest evidence would be requisite to make any sane man believe in the miracles by which Christianity is supported' (ibid., p. 86). But the sting comes in the tail, in words which were originally excised by Emma as 'too raw'. After the remark about the ease and slowness of the loss of belief, he added that he had 'never since doubted even for a single second that my conclusion was correct. I can indeed hardly see how anyone ought to wish Christianity to be true; for if so the plain language of the text seems to show that the men who do not believe, and this would include my Father, Brother and almost all my best friends, will be everlastingly punished. And this is a damnable doctrine' (ibid., p. 87).

The rejection of Christian claims to revelation belongs to the late 1830s and early '40s, the period of first working out his evolutionary theory. The question of the existence of a personal God did not arise until much later. When it did what had seemed the most powerful and decisive argument had lost its hold: 'The old argument of design . . . , as given by Paley, which formerly seemed to me so conclusive, fails, now that the law of natural selection has been discovered' (ibid., p. 87).

What in any case are we to think of a designer who must

be supposed to design and intend absolutely everything which exists or occurs in the universe? In 1860, in a letter to Asa Gray, Darwin put the straight question: 'An innocent and good man stands under a tree and is killed by a flash of lightning. Do you believe (and I really should like to hear) that God *designedly* killed this man?' (Darwin, F., I pp. 314–15): 'Does evil befall a city unless the Lord has done it?' (Amos 3: 6) A similar but much more disturbingly biological argument to the same effect is suggested in the final chapter of *The Origin of Species*, and developed in one of the finest of all sets of Gifford Lectures. In *Man on his Nature* Sir Charles Sherrington puts the same question with reference to the subtle and complex adaptation displayed in the life-cycles of two parasites, the liver-fluke and the Anopheles mosquito. Are all these things to be accounted the intended displays of a great but ghoulish Divine ingenuity? (Sherrington, pp. 366 ff.)

After reviewing two further considerations, which scarcely deserve to be dignified with the diploma description 'evidencing reasons for belief in God and immortality', Darwin confesses: 'Another source of conviction in the existence of God, connected with the reason and not with the feelings, impresses me as having much more weight. This follows from the extreme difficulty or rather impossibility of conceiving this immense and wonderful universe, . . . as the result of blind chance or necessity. When thus reflecting, I feel compelled to look to a First Cause having an intelligent mind in some degree analogous to that of man; and I deserve to be called a Theist' (Barlow, pp. 92–3).

But, although this conviction was at one time compelling, 'it has very gradually and with many fluctuations become weaker'. So the final upshot was an unhesitating rejection of all supposedly revealed religion, and for the rest agnosticism: 'I cannot pretend to throw the least light on such abstruse problems. The mystery of the beginning of all things is insoluble by us; and I for one must be content to remain an Agnostic' (ibid., p. 94).

(b) The first challenge to religious assumptions which is embodied or implied in *The Origin of Species* refers to doctrines of creation. The first point to appreciate about this challenge is, however, that the main shock of the impact of that book did not lie in any simple contradiction of statements in the first chapters of Genesis, literally interpreted. Many of these statements, thus literally interpreted, had been under scientific attack for sixty or more years. James Hutton's *Theory of the Earth*, published in 1795, had spoken most famously of how: 'We find no vestige of a beginning, no prospect of an end' (Hutton, I p. 200). We have already noticed the enormous influence of Lyell's *Principles of Geology*, which was for Darwin himself one of the formative books. This influence, as well as that of the much less scientifically reputable *Vestiges of Creation*, comes out very clear and hard in *In Memoriam*, a poem composed slowly over the years 1835–50 by Tennyson, a man who was to be described by T. H. Huxley as 'the first poet since Lucretius who has understood the drift of science' (Huxley, L., II p. 338). C. C. Gillispie's *Genesis and Geology*, subtitled *The Impact of Scientific Discoveries upon Religious Beliefs in the Decades before Darwin*, gives an excellent account of all these developments.

The uproar occasioned by the publication of *The Origin of Species* was not provoked primarily by the fact that its teaching – like that of the entire growing science of geology – was incompatible with the idea that the universe was constructed or created within the span of a single working week. That this was not then the chief stone of stumbling comes out most memorably when we reflect on the high spot of the notorious debate at the 1860 Oxford meeting of the British Association for the Advancement of Science. Darwin's Cambridge friend Henslow was in the chair. Huxley happened to be present and on the platform only thanks to the prior persuasions of Robert Chambers, whose *Vestiges of Creation* Huxley had in his review soundly caned. Bishop Samuel Wilberforce – known to the irreverent as 'Soapy Sam' – opened up in a light scoffing tone. He assured the audience that there was nothing in any idea of evolution: rock-pigeons, for instance, are what rock-pigeons have always been. But the climax, as one listener recorded, concerned what Darwin had not yet said about *The Descent of Man*: 'Then, turning to his antagonist with a smiling insolence, Wilberforce begged to know, was it through his grandfather or his grandmother that he claimed his descent from a monkey?' (Huxley, L., I pp. 265–6).

(i) How far the real crux is the same for Christian fundamentalists in the present century it is perhaps less easy to say. That it is, is suggested by the fact that the vastly publicized 1925 court confrontation is usually known as 'the Tennessee monkey trial' or 'the Scopes monkey trial'. The state of Tennessee had passed a law forbidding the teaching of evolutionary biology in its public schools. A young teacher John T. Scopes was prosecuted under this law. Both the original trial and the subsequent semi-documentary film had stars in their casts. William Jennings Bryan, sometime Democratic presidential candidate and famous as a silver-tongued orator, led for the prosecution, against the defence of Clarence Darrow – *Attorney for the Damned*. Thanks especially to the savage reporting of H. L. Mencken, 'the anthropoid rabble' of Tennessee were eventually shamed into overturning the conviction of Scopes on appeal, and soon after repealed the offending 'monkey law'. (In the movie *Inherit the Wind* Fredric March played the rather tragic part of Bryan, Spencer Tracy that of Darrow, and Gene Kelly, Mencken.)

Although this affair had the effects both of killing that law and discouraging potential imitators in other 'Bible belt' states, the battle was by no means over. It continues still. In the centenary celebrations of 1959, for instance, the Nobel Prize-winning American biologist H. J. Muller had to give his address under the sourly apt title 'One Hundred Years without Darwin is Enough'. The burden of that talk was that, whether because they had themselves been inadequately taught, or whether through fear of being condemned as controversial, less than half of America's high school teachers of biology were teaching evolution by natural selection. Many, too, of the largest and most respected publishers, fearing lest fundamentalist pressure in or on school boards should otherwise prevent the adoption of their products, continue still to produce school textbooks making no reference to either evolution or Darwin. (The London *Daily Telegraph* for 26 June 1982 carried a report from its local correspondent that the New York City School Board had recently decided to bring some counter-pressure by excluding three such textbooks from all the many schools under its control.)

There appears to have been within the USA during the last ten or fifteen years a revival of aggressively anti-evolutionary fundamentalist Christianity. The new tactic is not so much to try to keep the teaching of evolution out of the schools as to demand, both in schools and everywhere else, 'equal time' for what protagonists call 'creation-science'. This led in December 1981 to what the popular press described as 'Scopes II'. The state legislature of Arkansas had during the previous March, in what seems to have been a fit of absence of mind, passed a bill requiring that 'Public schools within this state shall give balanced treatment to creation-science and to evolution-science'.

The American Civil Liberties Union (ACLU) at once brought a case challenging the constitutionality of this law, mainly on the grounds that it violated the First Amendment separation of church and state. Scopes II was as drama not in the same class as Scopes I. But the ACLU's ultra-smart New York lawyers did call some very high powered and effective expert witnesses, in the shape of Stephen Jay Gould and the Canadian philosopher of biology Michael Ruse. And Ruse was afterwards able to boast that the verdict of the court could not have been more satisfactory had he written it himself.

The heart of the matter was that 'creation-science' is precisely not science, but a putative religious revelation. This is in effect admitted by Duane T. Gish in a passage of his *Evolution: The Fossils Say No!*, a passage which Ruse enjoyed reading out in court:

> By 'creation' we mean the bringing into being of the basic kinds of plants and animals by the process of sudden, or fiat, creation described in the first two chapters of *Genesis*. Here we find the creation by God of the plants and animals, each commanded to reproduce after its own kind using processes which were essentially instantaneous. We do not know how God created, what processes he used, for God used *processes which are not now operating anywhere in the natural universe*. This is why we refer to divine creation as special creation. We cannot discover by scientific investigations anything about the creative processes used by God. (Gish, pp. 24–5; italics supplied)

Only one thing needs to be added here. It is that Gish and his fellow obscurantists make much play with statements made by scientist clients of the demoralizing misconceptions discussed in Section 1, above. It would be tempting to reintroduce in a fresh context Julian Benda's fine phrase 'the treason of the clerks', were it not that here it is usually muddle-headedness or defeatism rather than calculating treason (Benda). Gish, for instance, at one point summons up support from the 1971 Everyman's Library edition of *The Origin of Species*, which makes a grotesque comparison: 'Belief in the theory of evolution is thus exactly parallel to belief in special creation – both are concepts which believers know to be true but neither, up to the present, has been capable of proof.' (This was one of several fresh introductions contributed at that time, after the appointment of a new General Editor, to classics which could in one way or another be upsetting to Roman Catholics.)

(ii) Once our consideration of creation begins to move beyond questions of the truth or falsity of certain particular statements in Genesis, in their most literal readings, there is one fundamental yet generally unfamiliar distinction to be made. This seems to have surfaced first in Aristotle's criticism of the cosmology in Plato's *Timaeus*. It is a distinction upon which St Thomas Aquinas was later to insist, even in face of murmuring accusations of heresy.

Most of us, when we hear the word 'creation', recall the first sentence of Genesis: 'In the beginning God created the heaven and the earth.' Some know, too, that Christian theologians have traditionally insisted that this was supposed to be, not any kind of transformation, but a creation out of nothing. The second verse is thus construed as describing not the materials for, but the product of, that original creation: 'And the earth was without form, and void; and darkness was upon the face of the deep. And the Spirit of God moved upon the face of the waters.'

Thomas, however, in a pamphlet *Concerning the Eternity of the Universe, against Murmurers* and elsewhere, maintained that 'there is no contradiction in affirming that a thing is created and also that it was never non-existent'. His point was that the absolute and constant dependence of every creature upon its creator, which is an essential of theism as opposed to deism, no more entails that all actual created things must have had a beginning, than it entails that any such things have had, or will have to have, an ending. That the universe is in this sense God's creation can be proved, Thomas held, by arguments of natural reason, and without appeal to the Christian or any other putative revelation. Without God as the sustaining first cause all things would – in the memorable phrase coined by Archbishop William Temple, a sometime British Primate – 'collapse into non-existence'. It is, presumably, to this conclusion that the first cause argument entertained in Darwin's *Autobiography* would be relevant.

That the universe was in fact created out of nothing 'in the beginning' was something which Thomas also believed. But this he believed as a matter of his faith in the supposedly revelatory teachings of the Catholic Church; just as he believed on the same grounds that there are some creatures – specifically, angels and human souls – from whom God will never in fact withdraw his existential support.

We now need a second distinction: between, on the one hand, the temporal creation of the entire universe; and, on the other hand, equally temporal particular and special creations within that universe. Whereas nothing in evolutionary biology has any direct relevance to the former, the latter would appear to be precluded – as the quotation from Duane Gish so wryly confirms – by the very nature of any project for an historical science, whether of biology or of geology or human history itself (Hume, 1748, X; and compare Flew, 1961, VIII). The science which is in a way relevant to the doctrine that the universe was created 'in the beginning' is cosmology. For a 'big bang' theory, as opposed to the hypothesis of Hoyle and Bondi, misleadingly characterized as 'continuous creation', would seem to be at least consistent with such a traditional conclusion, if not perhaps actually positive evidence for it.

Nowadays, when discussing such possible conflicts between religion and science, one has to take account of eirenic and somewhat self-consciously knowing suggestions that all those conflicts which must surely have arisen in the past sprang from nothing but ignorance and naïve muddle.

These suggestions should be firmly and briefly dismissed. For it is a sign not so much of high sophistication as of deeply sheltered ignorance to adopt as 'the standpoint of twentieth-century orthodoxy' the claim 'that – Science being descriptive, Religion normative – the apparent conflict between them sprang always from confusions' (Toulmin and Goodfield, pp. 226–7). Whatever may have been true of a few rare spirits in the past, whatever may be true of those many today who having lost their Christian faith cling nevertheless to the Christian name, it remains an egregious falsehood to suggest that the religion of the saints and of the fathers, of the popes and of the councils, either was or is altogether barren of descriptive intention. Even the fashionable so-called Liberation Theology, which has precious little to do with either God or liberation, still embraces the allegedly factual elements of a rather Third Worldly Marxism-Leninism (Flew and MacIntyre, Flew, 1966, and Flew, 1976b). And, furthermore, if and in so far as what Hume would have called 'any system of religion' really does entail both the literal truth of the first two chapters of Genesis and the special creation of species; then we can surely know, and not merely make an ingenuous claim to know, that that system as a system is in its descriptive assertions neither more nor less than false (Hume, 1748, p. 127).

(c) The second challenge to religious assumptions was the threat to the whole established system of rational apologetic. This system, developed throughout the eighteenth century, eventually found its classical expression in two works by William Paley: *Natural Theology* (1802) and *Evidences of Christianity* (1794). The latter came to be adopted as, and for a very long time remained, a set-book in the University of Cambridge. Darwin while an undergraduate certainly studied and was much impressed by the former also, which was at most recommended.

The first stage in this very systematic and rational system of apologetic consisted in a natural theology; an attempt, that is, to show by arguments of natural reason, as opposed to appeals to revelation, that the universe is God's creation. (A natural theology may also include putative proofs of the natural immortality of the human soul. But that large topic must be deferred to the next sub-section.) In the second stage, the hope was to supplement the somewhat sketchy religion of nature, which was the most that any natural theology could aspire to warrant, by the more abundant riches of divine self-revelation. But there have been, are and will be many rival claimants, all pretending to speak for Omnipotence. Christianity, or, more particularly, Jesus bar Joseph, has to be shown to be speaking with the only authentic voice. So this supreme and central claim was to be made out, primarily, by demonstrating that there is abundant historical evidence to establish the occurrence of the required miracles: the occurrence not only of endorsing miracles of a kind not vouchsafed to false and rival prophets; but the occurrence also of the constitutive miracle of miracles, the Easter Resurrection. For who but God himself could thus ostentatiously over-ride the laws of nature?

The Origin of Species threatened the systematic rational apologetic in two ways. In the first place, by eliminating such direct *ad hoc* divine activities from the biological field, it further diminished the credibility of all stories of miraculous interventions anywhere. In the second place, it undermined what has traditionally but misleadingly been called the Argument from Design, a form of argument which has always been the most popularly persuasive element in natural theology, and to which Paley devoted by far the greatest part of his book *Natural Theology*. The label 'Argument from Design' is inept, and we shall hereafter replace it by 'Argument to Design'.

The point is that these arguments are supposed to be not *a priori* deductions but arguments from experience. An Argument from Design, moving from admitted design to a designer or designers, would be deductive, and compulsively valid: to say that something has been designed while denying that there had been any designer or designers would be to contradict yourself. But an Argument to Design starts with something which looks as if it must have been designed, and then proceeds to argue that – all experience shows – such a thing could not have come into being without design, indeed without Design.

Paley's own most famous move is to begin with a watch: if from the observation of a watch we may infer the existence of a watchmaker, so surely, by parity of reasoning, from the observation of mechanisms so marvellous as the human eye we must infer the existence of a Great Designer, God? Paley repudiates as an alternative any suggestion

> that the eye, the animal to which it belongs, every other animal, every plant . . . are only so many out of the possible varieties and combinations of being which the lapse of infinite ages has brought into existence; that the present world is the relic of variety; millions of other bodily forms and other species having perished, being by the defect of their constitution incapable of preservation . . . Now there is no foundation whatever for this conjecture in anything which we observe in the works of nature. (Paley, I p. 32)

It is today fascinating to observe how many of the sorts of evidence which Darwin was to redeploy to support evolution by natural selection were offered in his *Natural Theology* by Paley to prove special creation and design. Homologies, for instance, satisfy Paley that 'we never come into the province of a different Creator'. Then all manner of examples of specialist adaptation to widely different environments reveal to him always and only the prevenient wisdom of the God who designed all these so various creatures, and placed each kind in one particular station to which it was most perfectly fitted.

Of course, if you think that you have other sufficient reasons for believing in the existence of that God, then you will be fully justified in continuing to discern in all these instances that same prevenient wisdom. You may then, like Asa Gray, accept Darwin's theory yet still thank him 'for his striking contributions to teleology . . . knowing well that he rejects design, while all the while he is bringing out the neatest illustrations of it'. You remain, however, if you do this, wide open to embarrassing reminders. For it is hard indeed while reflecting on the life and work of the ichneumonidae, the liver-flukes, or the Anopheles mosquito to hail their intending and designing creator as – in the curiously cosy words of one of the great Rationalists – 'the most lovable of substances' (Leibniz, p. 531).

What the acceptance of Darwin's theory does rule out is Paley's kind of Argument to Design. No doubt we can argue from a watch on a Caribbean beach to watchmakers, probably in Switzerland or Japan. We can do this because all our experience tells us that watches are, always and only, humanly manufactured products. But there neither is nor ever has been any parity of reasoning between this argument and the argument from eyes to an eye-designer. We have no experience of eye-manufacture, whether by men or God. As far as our no doubt wretchedly limited experience goes, eyes never are designed or made but rather – like Topsy herself – they 'just growed'.

This favourite version of the Argument to Design was first formidably criticized by Hume, as part of his comprehensive attack upon the whole established system of rational apologetic (Hume, 1748, X–XI and Hume, 1779; also compare Flew, 1961, VIII–X). Paley and his contemporaries, and indeed most of their more immediate successors, showed little sign of trying to come to terms with Hume's objections. It is, however, interesting to note that in the period of the M notebook Darwin at any rate was paying a lot of attention to Hume. There are several references to passages on the relations and similarities between human psychology and that of the brutes, mentions of both the *Dialogues concerning Natural Religion* and *The Natural History of Religion*, and a boost for the first *Inquiry* – 'Hume's essay on the Human Understanding well worth reading' (Gruber, p. 296; and compare pp. 348 and 351. None of these three items, by the way, is indexed).

To the earlier and, if you like, philosophical objections of the last paragraph but one, Darwin's work supplied massive scientific reinforcement. For this showed, if only in outline, how eyes and every other organ and organism might, indeed must have evolved without design; unlike the first watches which, along with their successors, we certainly know to have been both designed and manufactured. And, furthermore, if it is possible on these lines to provide a naturalistic account of the origins of all the species of living things, then there is no other and greater kind of marvel in the universe for which we can be forced to postulate supernatural design. (It is not without reason that Christians of the more old-time kind – Christians, you might say, of the Paley tendency – have seen evolutionary biology as the most dangerous science; nor that, in the opposite camp, enthusiastic members of the Rationalist Press Association have earned the nickname 'Darwin's Witnesses'!)

There is, however, another version of the Argument to Design which cannot be overthrown by present or promised scientific advances. While the popular version starts from particular marvels within the universe, marvels which are supposed to be naturalistically inexplicable, this is prepared to accept all such naturalistic explanations, and to take off from what is thereby revealed of the general regularities of the universe as a whole. Darwin's discoveries, like all other discoveries of order and regularity within the universe, become grist to the mill.

The second sort of Argument to Design can perhaps, without the exercise of any excess of sympathetic imagination, be dimly perceived in the fifth of the Five Ways of St Thomas (his five putative proofs):

> We observe that things without consciousness, such as physical bodies, operate with a purpose, as appears from their cooperating invariably, or almost so, in the same way in order to obtain the best result. Clearly then they reach this end by intention and not by chance. Things lacking knowledge move towards an end only when directed by someone who knows and understands, as an arrow by an archer. There is, consequently, an intelligent being who directs all natural things to their ends; and this being we call God. (Aquinas, I ii 2)

The nerve of this contention is that these teleological features, supposedly discerned throughout the universe, cannot be intrinsic to that universe but must be imposed upon it by an extraneous orderer 'which all men call God'.

To any argument of this second sort, Hume has again supplied the most decisive and devastating objection. It is an expression of what he liked to call – following Pierre Bayle – not Stratonian but Stratonician atheism: Strato of Lampsacus having been next but one in succession to Aristotle as director of the Lyceum. This objection is best put as questions:

> Whatever warrant could we have that the order which we discover in the Universe – which is necessarily the only one we either do or could know – is not, as it appears to be, intrinsic but

imposed? Have you yourself had any experience of other Universes which can be deployed to justify your conviction that their natural state, as it were, is quite different from that which you attribute now to the only Universe which any of us actually do or can know? (Flew, 1967, §§ 3.1–30 or Flew, 1971, VI 5)

There are some hints to suggest that Darwin took Hume's point, although he never removed a few idly polite allusions to the creator inserted into the second edition of *The Origin of Species*. For he always expressed gentle distaste for talk of 'the plan of the Creator', distaste on the grounds that 'nothing is thus added to our knowledge'. And in a letter dated 29 March 1863 to his close friend Joseph Hooker he confessed: 'I have long regretted that I truckled to public opinion, and used the Pentateuchal term "creation"' (Darwin, F., III p. 18).

(d) We come now at last to the third and most fundamental challenge, the challenge which Wilberforce thought to see off with his counter challenge to T. H. Huxley. (This Wilberforce, by the way, is not the same as the great emancipator; who could have made an honest claim to be, without lapsing into any arcane political code, a liberation theologian!) In *The Origin of Species* this most fundamental challenge is barely hinted: 'Light will be thrown on the origin of man and his history'; and 'Thus, from the war of nature, from famine and death, the most exalted object which we are capable of conceiving, namely the production of the higher animals, directly follows' (Darwin, C., 1859, pp. 458 and 459).

No one, however, who has seen any of the notebooks from the late 1830s and early '40s should be surprised: either that 'the distant future' in which 'I see open fields for far more important researches' turned out to begin in the immediately following decade, culminating in the publication of *The Descent of Man* (1871); or that it appears at this stage never to have crossed Darwin's mind that we might be 'capable of conceiving' as 'the most exalted object' the God of the Ontological Argument – 'a being greater than which no other can be conceived' (Flew, 1971, VI 2). For already in those first evolutionary investigations Darwin was exploring the implication that our species had, like every other, evolved. Indeed he was then and always interested in continuities between man and the brutes. During *The Voyage of the 'Beagle'* he had been struck by the extreme primitiveness of the people inhabiting Tierra del Fuego, and later it was Section IX of Hume's first *Inquiry* 'Of the Reason of Animals' on which he made a particular note (Gruber, p. 348).

First in 1917, as 'One of the Difficulties of Psycho-analysis', and then again in 1925 in 'Resistances to Psycho-analysis', Freud reflected 'how the psycho-analytic view of the relation of the conscious ego to an overpowering unconscious was a severe blow to human self-love'. Looking back at the earlier paper, he went on there: 'I described this as the *psychological* blow to men's narcissism, and compared it with the *biological* blow delivered by the theory of descent and the earlier *cosmological* blow aimed at it by the discovery of Copernicus' (Freud, V p. 173).

The Copernican revolution in astronomy was not only 'a severe blow to human self-love'. It was also an equally or more severe blow to Christianity and to other geocentric religions. For this revolution in astronomy knocked our earth from the centre of the universe, and opened up the possibility of other inhabited worlds; if not within the local parish of our small solar system then somewhere out among the 'fixed' stars (Koyré). This was a blow to Christianity not because, or not primarily because, some elements of pre-Copernican astronomy are, as well they might be, embedded in some stories in the Old Testament. It was a blow and, surely, a heavy body blow because the whole Christian system takes it absolutely for granted that the prime concern of the creator in his universe is with what is and is not done by human beings upon the central stage of earth. If there are other inhabited worlds, then what becomes of the basic, defining Christian doctrine of the Incarnation? Have there been no other falls and, if there have, no other incarnations? But then, if other incarnations, must the unity of the blessed and undivided Trinity break up, in order to make room for the additional members involved in those further incarnations?

The biological blow, with which alone we are concerned here, is again a blow not because, or not primarily because there are bits of pre-Darwinian biology embedded in sacred scripture. Of course, any Darwinian account of *The Descent of Man* must be irreconcilable with a fundamentalist reading of either of the two Genesis accounts of human origins (1:27–8 and 2: 7–8 and 18–25); just as findings of other sciences are similarly incompatible with literal readings of other bits of both that and other books of the Bible. But the really heavy blow, perhaps a knock-out blow, is to assumptions about the peculiar and special status of man. What is in truth thus threatened is not at all the idea that we are members of a unique and inevitably dominant species. That we surely are. What is threatened is the basic and essential Christian assumption that man is not, or is not wholly, a creature of flesh and blood; the assumption that we, unlike the brutes, have to anticipate, with hope or trepidation, an eternal destiny.

(i) Consider first what *The Descent of Man* does not imply. In its final paragraph we read: 'Man may be excused for feeling some pride at having risen . . . to the very summit of the organic scale . . . We must, however, acknowledge . . . that man with all his noble qualities, with sympathy which feels for the most debased, with benevolence which extends not only to other men but to the humblest living creature, with his god-like intellect which has penetrated into the movements and constitution of the solar system – with all these exalted powers – Man still bears in his bodily frame the indelible stamp of his lowly origin' (Darwin, 1871, pp. 946–7).

What most certainly does not follow from anything discovered or said by Darwin is that there are no important differences between our species and its primate forebears, that people are really, or merely, or nothing but apes; or, indeed, anything else but people. To say that this evolved from that, since it presupposes that this is not identical with that, is precisely not to say that, really or ultimately, this must be that. It is no kind of evolutionary, or scientific, or any other sort of insight, to maintain that oaks are really – even in the last analysis – nothing but acorns.

Such reductionist misinterpretation can be found in serious scholarship as well as in works of catchguinea popularization. It was, for instance, a lapse when the distinguished American author of *Darwin and the Darwinian Revolution* wrote of his suggestions in *The Descent of Man*: '. . . as he earlier reduced language to the grunts and growls of a dog, he now contrived to reduce religion to the lick of the dog's tongue and the wagging of his tail' (Himmelfarb, p. 307). But in such British best-sellers as *The Naked Ape* and *The Human Zoo* similar reductions of the civilized to the savage and the human to the subhuman are not momentary aberrations but the chief stock-in-trade. Desmond Morris as the popularizer of science steps forward to reveal what he and zoologist colleagues have supposedly discovered: that man today is, when the chips are down, 'a primitive tribal hunter, masquerading as a civilized, super-tribal citizen' (Morris, 1970, p. 248). Generally the ape instincts overshadow everything learnt: thus, in listing conditions of inter-group violence among people, Morris explains that he has 'deliberately omitted . . . the development of different ideologies. As a zoologist viewing man as an animal, I find it hard to take such differences seriously' (ibid., pp. 47–8; and compare Flew, 1978, pp. 29–33).

Not only is it both unevolutionary and absurd to present evolved products as themselves being no more or other than what they evolved from, it is also preposterous to offer as a trophy of biological enlightenment a systematic refusal to attend to the respects in which our species differs from all the rest. To anyone who truly is thinking zoologically, the first peculiarities to leap to mind must surely be the far-extended period between birth and maturity, the incomparable capacity for learning, and the prominence of learned as opposed to instinctual behaviour. This unparalleled capacity for learning, together with its instrument and expression, developed language, provides our species with a serviceable substitute for the inheritance of acquired characteristics. (I must say here in my text, and not in an unread note, that those last thoughts are borrowed from Julian Huxley's *Essays of a Biologist*: which is as much a model of how the subject ought to be presented to the laity as the works of Desmond Morris are object-lessons of the opposite.)

A main support for the misguided reductionism discussed in the previous four paragraphs is the notion that differences of degree cannot be of essential and ultimate importance. Certainly Darwin and any Darwinian is committed to the view that the differences between species, like those between varieties, are one and all differences of degree; in the sense that there is, has been, and could have been, a spectrum of resembling cases stretching without interruption between any extremes of difference – a spectral series across which any line of division which we may draw cannot but be more or less arbitrary. This, as was brought out in Section 1 of the present Chapter II, is a clear consequence of the theory: 'When the views entertained in this volume . . . are generally admitted . . . The endless disputes whether or not some fifty species of British brambles are true species will cease. Systematists will have only to decide (not that this will be easy) whether any form be sufficiently constant and distinct from other forms, to be capable of definition; and, if definable, whether the differences be sufficiently important to deserve a specific name . . . Hereafter we shall be compelled to acknowledge that the only distinction between species and well-marked varieties is, that the latter are known, or believed to be connected at the present day by intermediate gradations, whereas species were formerly thus connected' (Darwin, 1859, p. 455).

This general conclusion Darwin insisted on applying to our own species too. He was appalled to learn that Alfred Russel Wallace – who in his later years got both spiritualism, and a sort of religion, and phrenology – wanted to make an exception in our favour (Darwin, F., III p. 116). If Darwin is right, then there is not and never has been any sharp, undisputatious, naturally given, antecedently correct dividing line between the human and the non-human or the not-yet human. Nor is there any guarantee that in every perplexing marginal case which we may in the future meet we shall be able – if once given all relevant facts – to return an unequivocally true or false answer to the question: 'Is this a human being or is it not?' This is a disturbing conclusion, which many may have difficulty in accepting. Others may find it hard to grasp even that it is, whether true or false, a necessary consequence of Darwinian theory (Warren, in Flew and Warren). It is a conclusion categorically rejected by St Augustine and by other great doctors of the Christian Church: the brutes, he insisted, all homologies notwithstanding, 'are not related to us by a common nature' (Augustine, 1966, p. 91).

Nevertheless, although any Darwinian is necessarily committed to rejecting St Augustine's denial, this rejection does not at the same time require, on the part of the Darwinian, corresponding denials of any uniqueness in man or of any importance in all the various relevant differences of degree. In the first place, although Man was no doubt 'formerly . . . connected' with primate ancestors 'by intermediate gradations', this seems 'at the present day' to be the case no longer. In the second place, and much more important, it is as preposterous as it is common to dismiss all differences of degree, however great and in whatever direction, as nothing but mere differences of degree. For almost if not quite all those differences which are of importance for living are, in the sense just now expounded, differences of degree. Can there really be anyone so reckless, so doctrinally infatuated, as to dismiss as trivial all differences between sanity and insanity, youth and age, riches and poverty, a free society and one in which whatever is not forbidden is compulsory? (Flew, 1975, §§ 7.13–24)

(ii) St Augustine, of course, did not rest his case upon this popular misconception – the essential triviality, that is, of all differences of degree. Instead, he Platonized. Both in his pamphlet *de Quantitate Animae* (On the Greatness of the Soul) and elsewhere he argued that we are not, or are not exclusively or mainly, creatures of flesh and blood. The essential and controlling element in all specimens of the species homo sapiens, the soul, is instead, like God himself, incorporeal. This ghostly and incorporeal soul, which survives the death and dissolution of the body, is, after re-equipment with a new and improved body, destined for an eternity of bliss; or, in the vast majority of cases, torment. It is to this view of the nature and destiny of man that the Roman Church held, and continues to hold.

So when in his epoch-making *Discourse on the Method* (1637) Descartes sketched what is sometimes called the Cartesian Compromise, he provided for a ghost as well as for a machine (Ryle). Descartes was a founding father both of modern science and of modern philosophy. But he was also Jesuit-educated and, at least to all appearances, a faithful son of the church. While, therefore, in his view 'the body is regarded as a machine which, having been made by the hands of God, is incomparably better arranged, and possesses in itself movements which are much more admirable, than any of those which can be invented by men', nevertheless the essential Descartes is a kind of ghost, an incorporeal subject of experience intimately integrated with, yet only temporarily lodged in, that bodily machine. Descartes later claims to have 'described . . . the rational soul and shown that it could not in any way be derived from the power of matter . . . but that it must be expressly created . . . And . . . inasmuch as we observe no other causes capable of destroying it, we are naturally inclined to judge that it is immortal' (Descartes, I pp. 116 and 117–18).

In what we must now call the First Vatican Council (1869–70) the Roman Church laid down two dogmatic canons insisting on both the possibility and the validity of both stages in the systematic rational apologetic discussed in sub-section 2(c), above:

> If anyone shall say that the one and true God, our creator and Lord, cannot through those things which are made be known for certain by the natural light of human reason, let them be cast out [anathema; and] If anyone shall say that miracles cannot happen . . . or that we can never know for certain of their occurrence nor properly prove the divine derivation of the Christian religion thereby, let them be cast out. (Denzinger, §§ 1806 and 1813)

The Cartesian Compromise was in 1953 similarly endorsed by Pope Pius XII, in the encyclical *Humani Generis*: '. . . the teaching of the Church leaves the doctrine of evolution an open question, as long as it confines its speculations to the development, from other living matter already in existence, of the human body'; nevertheless 'That souls are immediately created by God is a view which the Catholic faith imposes on us' (ibid., § 3027).

This compromise was no doubt all very well in the days of Descartes, for whom it is named, and for a couple of centuries thereafter. Yet to continue to demand that it has to be maintained more than a century after Darwin should be quite unacceptable. For it requires us to allow: not only that all individual souls are specially created, every such creation involving one of the miracles previously believed to initiate species; but also that human behaviour and experience are not to be explained in terms of the internal ongoings of the flesh and blood behaving organism, and in particular those in its central nervous system.

The acceptance of Darwinism must give much greater urgency to the question whether human souls are supposed to be scientifically detectable. Neither of the only two possible alternative answers holds out great attractions to the religious apologist. If the reply is that souls are at least in principle discernible in our experience, then the tough-minded are going to ask why they have not yet been detected and isolated, and when we may expect to hear from the science front news of this most spectacular of investigatory achievements. If, in order to avoid the embarrassment of such jeering queries from intellectual hardpersons, the reply comes in the opposite sense; then the appropriate yet no less damaging counter is philosophical rather than scientific. For what significance is now being given to talk of souls, if there is to be no conceivable means of identifying their presence or absence? And, furthermore, how could such supposedly conceivable entities even in principle either be individuated one from another or individually reidentified through time? (Flew, 1964, Introduction and Flew, 1976b, III)

III

SOCIAL SCIENCE, EVOLUTIONARY BIOLOGY AND SOCIOBIOLOGY

Very few attempts have been made to carry over conceptions derived from sociology into biology.

Julian Huxley, *Essays of a Biologist*, pp. 70–1

In October 1838, that is, fifteen months after I had begun my systematic enquiry, I happened to read for amusement Malthus *On Population*, and, being well prepared to appreciate the struggle for existence which everywhere goes on from long-continued observation of the habits of animals and plants, it at once struck me that under these circumstances favourable variations would tend to be preserved, and unfavourable ones to be destroyed. The result of this would be the formation of new species. Here, then, I had at last got a theory by which to work . . .

Charles Darwin, *Autobiography*, p. 57

Darwin grafted Adam Smith upon nature to establish his theory of natural selection . . .

Stephen Jay Gould, *Ever since Darwin*, p. 100

Darwin's book is very important and it suits me well that it supports the class struggle in history from the point of view of natural science. One has, of course, to put up with the crude English method of discourse.

Karl Marx, letter to Ferdinand Lassalle (16 January 1861)

The four quotations above are sufficient to show the need for some attempt to sort out the relations, or the lack of relations, between evolutionary biology and the social sciences. This attempt will consist of four sections. Section 1 will consider the influence of Malthus on Darwin; and how Darwin adapted for application to the non-human living world part of a theoretical scheme originally modelled on Newtonian classical mechanics, and only later and not thoroughly altered to make room for the realities of choice. Section 2 will ask whether there was any direct or indirect contribution from Adam Smith and the other Scottish founding fathers of social science; especially through their recognition that what look like products of intelligent design may be, and sometimes must have been, the unintended products of intended actions and their interactions. Section 3 will ignore the unrewarding question of possible Darwinian influence on Marx, investigating instead two other issues: whether Darwin's theory does in truth supply any sort of support for the revolutionary revelations of *The Communist Manifesto*; and whether we can accept the claim that Marx was, in his different field, a scientist of the same stature as Darwin. At the graveside Engels said, it will be remembered, that 'Just as Darwin discovered the law of development of organic nature, so Marx discovered the law of development of human history.' Finally, in Section 4, we shall come to the claims and ambitions of sociobiology, that Neo-Darwinian candidate discipline whose practitioners aspire to re-establish authentically scientific social sciences upon properly biological foundations.

1. MALTHUS: POWERS, CHECKS AND CHOICE

That momentary oversight by Julian Huxley notwithstanding, it is one of the most remarkable and most often remarked facts of the history of ideas that it was reading T. R. Malthus's *An Essay on the Principle of Population* which provided the crucial stimulus first to Darwin himself and later to Alfred Russel Wallace. In the case of Darwin, though not of Wallace, later recollections are abundantly confirmed by contemporary private notes (Flew, 1970, pp. 49–51, and further references there provided). What perhaps excuses Huxley's neglect of this most important case of carrying over 'conceptions derived from sociology into biology' is the fact that Malthus himself was insisting that a principle which applies both to plants and to the brutes must apply to human populations also. The best statement is in *A Summary View of the Principle of Population*, which contains the greater part of the article 'Population' written by Malthus for the 1824 Supplement to the *Encyclopaedia Britannica*:

> In taking a view of animated nature, we cannot fail to be struck with a prodigious power of increase in plants and animals . . . Elevated as man is above all other animals by his intellectual faculties, it is not to be supposed that the physical laws to which he is subjected should be essentially different from those which are observed to prevail in other parts of animated nature: . . . all animals, according to the known laws by which they are produced, must have a capacity of increasing in a geometrical progression. (Malthus, 1824, pp. 119, 121–2 and 123)

It is precisely this principle, a principle of all biology both human and non-human, which was to lead both Darwin and Wallace to recognize natural selection as the main if not the only mechanism of evolution. In seeing this, Darwin at the same time recognizes that the Malthusian principle of population cannot apply in exactly the same way to human beings: 'A struggle for existence inevitably follows from the high rate at which all organic beings tend to increase . . . as more individuals are produced than can possibly survive, there must in every case be a struggle for existence, either one individual with another of the same species, or with the individuals of distinct species, or with the physical conditions of life. This is the doctrine of Malthus applied with manifold force to the whole animal and vegetable kingdom; for in this case there can be no artificial increase of food and no prudential restraint from marriage' (Darwin, C., 1859, pp. 116–17).

(a) The whole doctrine of Malthus, of course, was much more extensive than this. His insistence upon the power or capacity in human populations 'of increasing in a geometrical progression' was only the first principle of the conceptual scheme which guided and structured all his enormously influential work in this area. Apart from shorter pieces, such as the *Britannica* article mentioned in the last paragraph but one, there were two books which were originally published, and hence are still by all librarians treated, as, respectively, the first and the second and later editions of *An Essay on the Principle of Population*. The first of these, now known by Malthus scholars as the *First Essay*, was a swingeing polemical pamphlet. Against all comers, including and perhaps especially his own fond but starry-eyed father, Malthus argued that all present and future utopian schemes for the construction of ideal societies must be ruined by the inexorable pressures of population (Malthus, 1798). The *Second Essay* was very different, and nearly four times as long (Malthus, 1802). It was a treatise on population, incorporating the results of the considerable research done by Malthus in the four intervening years. 'In its present shape,' as he said in the Preface, 'it may be considered as a new work.' It may be, and it should. Here and in all subsequent statements, Malthus made room for the possible and desirable check of Moral Restraint, very narrowly defined by him as restraint from, although apparently not within, marriage, and specifically excluding all contraception or abortion as vicious and immoral. This admission, he was happy to remark, allowed him 'to soften some of the harshest conclusions of the *First Essay*'.

It did not, however, soften the tempers of all those infuriated by the uncomfortable yet inescapable remaining conclusion – that there can be no realistic prospect of any long-term improvement in the human condition without some sort of effective check on population. It is, surely, discomfort on this count which not wholly but in large part explains why the work of Malthus has been and is so often misrepresented, and Malthus himself put down as some kind of moral and intellectual delinquent (Flew, 1970). Gavin de Beer, for instance, in what remains the standard biography of Charles Darwin goes out of his way to abuse Malthus for producing only 'special pleading masquerading as science' (de Beer, p. 20; and compare pp. 99–101).

The most apoplectic reaction and, because of their colossal influence as, in Bertrand Russell's phrase, the 'founders of the crusading new Islam of the twentieth century', the most unfortunate, was that of Marx and Engels. In his *Outlines of a Critique of Political Economy* Engels railed at the 'sham philanthropy' which 'produced the Malthusian population theory – the crudest, most barbarous theory which ever existed, a system of despair which struck down all those beautiful phrases about love of neighbour and world citizenship'. Later he asks, starting with words which sound more than a little incongruous on the lips of an atheist: 'Am I to go on any longer elaborating this vile, infamous theory, this revolting blasphemy against nature and against mankind? Am I to pursue its consequences any further? Here at last we have the immorality of the economist brought to its highest pitch' (Engels, 1844, pp. 199 and 219). Marx follows Engels in similarly spluttering invective: 'The hatred of the English working class against Malthus – the "mountebank-person" as Cobbett rudely calls him – is therefore entirely justified. The people were right here in sensing instinctively that they were confronted not with a man of science but with a bought advocate, a pleader on behalf of their enemies, a shameless sycophant of the ruling classes' (Meek, p. 123).

If we are ever to get a grasp on the relations or lack of relations between theories in the social and theories in the biological sciences, then we shall have, I fear, to defy the wrath of Engels. We shall have to go on, at least a little longer, 'elaborating this vile, infamous theory, this revolting blasphemy against nature and against mankind'. The first things to appreciate are: that the organizing conceptual scheme of Malthus was with intent modelled upon that of classical mechanics; and that he himself was well grounded both in pure mathematics and physics – unlike some of the many others who have aspired to prepare the way for, or even to be, the expected new Newton of the human sciences. He had graduated at Cambridge in 1788 as ninth Wrangler; which, being translated, means that he had been awarded the ninth best First in the honours school of mathematics and mathematical physics.

The function, within his scheme, of the Principle of Population was the same as that, within the inspiration of that scheme, of the First Law of Motion. Neither is supposed to describe what is always and everywhere observed to happen. On the contrary: it is precisely and only because things are observed not in fact to be thus and thus that the heuristically seminal question arises: 'Why not? What are the various checks which prevent populations multiplying up to the theoretically possible limit?'

Because, like so many other workers in the field of the

social sciences both then and since, Malthus was as much or more interested in practical policy, he proposed to classify possible checks in two entirely different ways. One was value–neutral: in that classification, ultimately, the two mutually exclusive categories were 'positive' and 'preventive'; the former embracing all causes of death, and the latter everything preventing the occurrence of a possible birth. The other kind of classification was, very heavily, value-loaded. Here in the *First Essay* there were again only two categories, supposedly exhaustive yet not – surely? – mutually exclusive. It was to these two forbidding categories of vice and misery that in the *Second Essay* he added a perhaps not to everyone substantially more appealing third, 'Moral Restraint'.

To complete the theoretical structure Malthus makes the point that the values of the various possible checks do not vary entirely independently: 'The sum of all the positive and preventive checks, taken together, forms undoubtedly the immediate cause which represses population . . . we can certainly draw no safe conclusion from the contemplation of two or three of these checks taken by themselves because it so frequently happens that the excess of one check is balanced by the defect of some other' (Malthus, 1802, I p. 256). Although his general statements about the relations between the various checks considered as variables are usually, like this one, curiously weak, his particular arguments again and again depend on the subsistence of far stronger connections. Thus, in the *First Essay* he remarks that the failure of Richard Price, after supposing that all the checks other than famine were removed, to draw 'the obvious and necessary inference that an unchecked population would increase beyond comparison, faster than the earth, by the best directed exertions of man, could produce food for its support' was 'as astonishing, as if he had resisted the conclusion of one of the plainest propositions of Euclid' (Malthus, 1798, p. 197). Again, in the *Second Essay*, Malthus quotes with approval the remark of a Jesuit missionary: 'if famine did not from time to time, thin the immense number of inhabitants which China contains, it would be impossible for her to live in peace.' Most significant of all, the whole force of the argument for Moral Restraint lies in the contention that this check must be substituted for those others which Malthus classed as species of vice or misery.

(b) It is, and Malthus always sees and insists on the point, very like classical mechanics. Thus in Book I of Newton's *Principia* the First Law of Motion runs: 'Every body continues in its state of rest or of uniform motion in a right line unless it is compelled to change that state by forces pressed upon it.' Since in actual fact all bodies are in motion relative to at least some others, and since this motion never continues for long in a right [straight] line, the questions arise: 'Why do bodies not continue in a state of rest or of uniform motion in a right line? What forces operate to prevent this, and how?'

In the Appendix added to the *Second Essay* in 1817 Malthus defends his talk of a natural tendency, which in fact is always to a greater or lesser extent checked by counteracting forces, by appealing to the practice 'of the natural philosopher . . . observing the different velocities and ranges of projectiles passing through resisting media of different densities'. He complains that he cannot 'see why the moral and political philosopher should proceed upon principles so totally opposite'.

It is not only very like but also deliberately modelled on the Newtonian paradigm of modern physical science. The trouble is that it is too like, the modelling is too faithful. For the human sciences have always to take account of a reality which does not fall within the scope of the purely physical, the reality of choice. There is, therefore, an enormous and crucial difference: between, on the one hand, the power of a human population to multiply if 'left to exert itself with perfect freedom'; and, on the other hand, the kind of natural power described by the First Law of Motion. It is a difference of which Malthus began to take explicit account when he admitted the possibility of Moral Restraint. But this admission demands theoretical adjustments much more drastic and pervasive than he ever recognized to be required. There is also a difference which was pointed out by contemporary critics, and somewhat half-heartedly accepted by Malthus, between two related senses of the word 'tendency'.

(i) Consider again some statements quoted earlier:

In taking a view of animated nature, we cannot fail to be struck with a prodigious power of increase in plants and animals ... Elevated as man is above all other animals by his intellectual faculties, it is not to be supposed that the physical laws to which he is subjected should be essentially different from those which are observed to prevail in other parts of animated nature ... all animals, according to the known laws by which they are produced, must have a capacity of increasing in a geometrical progression.

Faced by this attempt at a natural science in the peculiar area of human action, we have to make and to insist upon a fundamental distinction between two senses of the word 'power'. In one sense, the only sense in which the word can be applied to inanimate objects and to most of animate nature, a power simply is a disposition to behave in such and such a way, given that such and such preconditions are satisfied. Thus we might say that the bomb ('the nuclear device') dropped at Nagasaki possessed an explosive power equivalent to that of so many tons of TNT, or that full-weight nylon climbing rope has a breaking strain of (a power to hold up to) 4,500 pounds. Let us, for future ready reference, label this 'power (physical)'. In another sense, the sense in which the word is typically applied to people, and perhaps to people only, a power is an ability at will either to do or to abstain from doing whatever it may be. Thus we might say that in his heyday J. V. Stalin had the power of life and death over every subject of the Soviet Empire, or that a fertile pair of people of opposite sexes has the power to start a baby. Call this 'power (personal)'.

This second sense of the word 'power' essentially involves the possibility of choice and, of course, wherever that possibility is present it is present inescapably. To choose not to choose is still, necessarily, to choose (Locke, II xxi 23 p. 245). There is much more to be said about this sort of power, and about what statements that someone possesses it do and do not entail. But here and now it must suffice to make a brief quotation from the chapter 'Of Power' in Locke's *Essay Concerning Human Understanding*. The sentences chosen both explain what is meant by having or not having the power to do or to refrain from doing something at will, and indicate the utterly familiar differences in terms of which this distinction is and has to be defined. They thus bring out at one and the same time: both what is meant by assertions that he or she has the power to do this or that; and that it is altogether out of the question to deny, on any grounds whatsoever, that anyone who is not unconscious and totally paralysed possesses at least some such powers.

Locke speaks with his customary common sense and caution. Yet we must give warning that – like so many successors still – he confuses questions about what it is to be an agent with questions about what it is to be a free agent. For even the person acting not freely but under extreme compulsion is nevertheless an agent who, in the fundamental sense which Locke is about to explain, could do otherwise. His situation is, therefore, that of the businessman who receives from the Godfather 'an offer which he cannot refuse'; rather than that of the errant mafioso who, gunned down from behind, collapses into a pool of his own blood. In the most literal if not in the more idiomatic sense the first Mafia victim did have a choice – was it to be his signature or his brains on the extortion document? But the renegade mafioso did not: he had been an agent; but, in that moment of sudden death, ceased to be. (It is as well also to say that the Latin in the quotation below translates into 'St Vitus' dance'.)

Section 7 of the chapter 'Of Power' reads. 'Everyone, I think, finds in himself a power to begin or forbear, continue or put an end to several actions in himself. From the consideration of the extent of this power . . . which everyone finds in himself, arise the ideas of liberty and necessity.' Then in Section 11 he continues:

> We have instances enough, and often more than enough in our own bodies. A Man's heart beats, and the blood circulates, which 'tis not in his power . . . to stop; and therefore in respect of these motions, where rest depends not on his choice . . . he is not a *free Agent*. Convulsive motions agitate his legs, so that though he *wills* it never so much, he cannot . . . stop their motion (as in that odd disease called chorea Sancti Viti) but he is perpetually dancing: He is . . . under as much Necessity of moving, as a Stone that falls or a Tennis-ball struck with a Racket. (Locke, pp. 237 and 239)

(ii) So much for power and necessity. The relevant distinction between two senses of the word 'tendency' was well made by Archbishop Whately in 1832, in the ninth of his *Lectures on Political Economy*: in one a tendency to produce something is a cause which, operating unimpeded, would produce it; in the other to speak of a tendency to produce something is to imply that that result is in fact likely to occur. Malthus, misled perhaps by his favourite physical paradigm, seems to have slipped without distinction from the first interpretation, which comes easily to the theoretical natural scientist, to the second, which belongs rather to the discussion of practical human affairs. It is as if one were to argue that because the First Law of Motion is, in the first sense, a law of tendency, it must therefore follow that it is probable or certain that everything, in the second sense, tends to remain at rest or to move uniformly in a right line. In a similar way, especially but not only in the *First Essay*, Malthus was inclined to construe the multiplicative power of human populations as a natural force rather than as, what it is, a human power.

It is these confusions which mainly determine the gloomy conclusions actually drawn in the *First Essay*; as well perhaps as those even gloomier conclusions which Malthus surely should have drawn, and which he is still so often falsely accused of having not only drawn but relished. In particular, they surely do imply what we may mischievously christen Parkinson's Law of Population – the doctrine that always and everywhere human populations must press hard up against whatever resources are available for their support. It was this doctrine which Malthus employed to shatter all utopian dreams of universal egalitarian abundance. For if you undertake to provide generously for everyone, and for their families too, and if Parkinson's Law of Population is true, then your resources must be rapidly overwhelmed by the torrent of births: effective demand for welfare beneficiaries is, as has been suggested by certain dry economists among our own contemporaries, bound to produce an increasing and more than adequate supply! Not for nothing was the *First Essay* published as *An Essay on the Principle of Population as it affects the Future Improvements of Society, with Remarks on the Speculations of Mr Godwin, M. Condercet, and other Writers.*

It must, surely, have been precisely and only because Malthus was throughout this *First Essay* construing his Principle of Population on a Newtonian model, and in a way which made it imply population Parkinsonism, that he became unable to recognize any possibilities of voluntary control, any possibilities of individual or collective policies for the inhibition of this mighty and menacing power of multiplication. Certainly it has to be seen as a fact crying out for explanation that so able and so concerned a writer should even for a short time have failed to recognize the possibility of any form of Moral Restraint in this area. For Malthus was never committed to any general doctrine of hard determinism, requiring him to deny the possibility of choice in this and all other particular cases. On the contrary: even in the *First Essay* itself he has a bit to say about the wrongness of choosing vicious rather than virtuous alternatives. So, I am proposing, the explanation of his temporary, local and limited psychic blindness has to be found in his understandable yet unfortunate infatuation with his Newtonian model; a model which must indeed warrant the moral of a helpless acceptance of population Parkinsonism.

2. MAKING VISIBLE THE INVISIBLE HANDS

In the case of Malthus we have Darwin's own statements of what he owed, and it was necessary only to put that debt into the context of the whole Malthusian conceptual scheme. In doing this we had occasion to bring out some interesting oddities in the movement of ideas: in particular that, in order to make Malthus relevant to biology below the human level, Darwin had to discount the possibility of Moral Restraint; the very possibility which Malthus himself had had to admit in order to adapt his Newtonian scheme to the fundamentally different field of human demography.

In the matter of possible direct or indirect influence from the ideas of Adam Smith we have, I believe, no sort of acknowledgement in any of the published works of Darwin, and only somewhat indirect indications in the surviving notebooks and other private papers. On the other hand one most important encouragement which he might have derived from Smith, or indeed from any of several other major figures of the late eighteenth-century Edinburgh Enlightenment, would have come from something quite distinctive of and peculiar to the human sciences. That, as we have seen, is certainly not the case with what Darwin did in fact get from Malthus. The point is that in case after case all these great Scots had been arguing that many institutions and constitutions which might look as if they must have been products of inspired individual or consciously co-operating collective design in fact were, and sometimes could not not have been, unintended results of intended actions (Hayek, 1967).

(a) The first suggestion to consider is that made first perhaps by Marx – that Darwin saw in subhuman nature an analogue to the pluralist and competitive market economy of Victorian England. There can, I think, be no doubt but that Darwin did see similarities. Well read, generally well informed, and with the closest of personal ties with the Wedgwoods, he could scarcely have failed so to do. So the question is: 'So what? What do you want to make of the facts that there are similarities, and that Darwin saw them?'

No one, surely, wants to deny that these similarities did obtain, to deny that there really is a struggle for existence at the subhuman level, to deny that more offspring are produced than either will or can survive to reproduce? No one, surely, wants to maintain that Darwin was falsely projecting his picture of what are in truth exclusively human struggles out on to what is in reality an altogether peaceable kingdom? The popularizers who remark both the similarities and Darwin's awareness of them appear usually to be concerned only to get in those digs against capitalism and against the Victorians which are felt to be necessary in order to maintain their own status as 'progressive people' and, above all, 'not in the least right-wing'. Certainly to go further than this would be for such socializing trendies imprudent, as well as mistaken. For to suggest, as some sociologists of belief have done, that Darwin's theory could as a matter of fact only have been produced in the sort of society in which it actually was produced is to concede yet another comparative advantage to competitive capitalism as against monopoly socialism.

(b) Recent writers have drawn attention to the evidence that Darwin was in the late 1830s attending to the ideas not only of Hume but also of Adam Smith (Gruber, Schweber, Vorzimmer). There is, for instance, a reference in the M notebook to Dugald Stewart's 'Account of the Life and Writings of Adam Smith, Ll. D.', an account which is printed as a very long Appendix to the posthumous *Adam Smith: Essays on Philosophical Subjects* (Gruber, p. 286; and compare Wightman). We also know that Darwin was in this period reading 'for amusement' the novels of Harriet Martineau (Vorzimmer). These best-selling 'romances', with their strong simple plots, also included conscientiously full commentaries upon the ideas of Smith, Malthus and Ricardo.

It is, therefore, relevant now to quote again that often quoted yet much more often abused and misunderstood passage from *The Wealth of Nations*:

> But it is only for the sake of profit, that any man employs a capital in the support of industry . . . As every individual, therefore, endeavours as much as he can . . . to employ his capital . . . that its produce may be of the greatest value; every individual necessarily labours to render the annual revenue of the society as great as he can. He generally, indeed, neither intends to promote the public interest, nor knows how much he is promoting it . . . by directing . . . industry in such a manner as its produce may be of the greatest value, he intends only his own gain, and he is in this as in many other cases, led by an invisible hand to promote an end which was no part of his intentions. Nor is it always the worse for the society that it was no part of it. By pursuing his own interest he frequently promotes that of the society more effectually than when he really intends to promote it. (Smith, IV ii pp. 455–6)

(i) For us the first and crucial point to seize concerns that famous or, if you insist, notorious invisible hand: '. . . he intends only his own gain and he is in this as in many other cases, led by an invisible hand to promote an end which was no part of his intentions.' It is on his attention, 'in this as in many other cases', to such unintended consequences of intended action that Smith's claim to have been one of the founding fathers of the social sciences is most properly founded.

To understand that claim is to realize how totally wrong it is to construe Smith's invisible hand as an instrument of providential direction. To do this would be as preposterous as to interpret Darwin's natural selection as being really supernatural selection. For Smith's invisible hand is no more a hand directed by a rational, intending and designing owner than Darwin's natural selection is selection by supernatural and providential intelligence. In the proverbial nutshell: the invisible hand is no more a hand than natural selection is selection. Both Smith and Darwin are showing how something which one might be very tempted to put down to design could and indeed must come about: in the one case without direction, in that direction; and in the other without any direction at all. By uncovering the mechanisms operative in the two cases they both made supernatural intervention as an explanation superfluous.

Nor would it be right to accuse Smith of assuming that the effects of the operation of all such unplanned and unintended social mechanisms are always and exclusively happy. For there was at least one other known to him, to the costs of which he was most sensitive:

> The division of labour is not originally the effect of any human wisdom, which foresees and intends the general opulence to which it gives occasion. It is the necessary though very slow and gradual consequence of a certain propensity in human nature which has in view no such extensive utility: the propensity to truck, barter, and exchange one thing with another. (ibid., I ii p. 25)

But it is actually Smith himself whom in *Capital* Marx cites on the lamentably dehumanizing consequences of the division of labour; although Marx can, of course, provide no clue on how the proletarian apocalypse promised in *The Communist Manifesto* is to avoid the progressive intensification of such divisions (Marx, 1867, I p. 362).

Perhaps the most striking and important example of a social mechanism with entirely unhappy effects is what Garrett Hardin has labelled 'The Tragedy of the Commons'. Where without restrictions of private property access is common to several, those sharing that access will all and rationally be inclined to make the most use they can of any resource, which will therefore tend to be wastefully and rapidly exhausted or destroyed: a result universally unintended and unwanted. In our contemporary world one appalling token of this type is the ruin of Sahel. There, as Kurt Waldheim has warned, 'the encroachment of the desert threatens to wipe out four or five African countries from the map'. Certainly other causes, such as protracted drought, have exacerbated the problem. But the basic trouble is that on unenclosed land no one has an individual interest in doing what stops, or not doing what starts, desertification. Notoriously, everyone's business is no one's. Whereas, as Arthur Young, one of Adam Smith's younger contemporaries and the first agricultural journalist, was wont to say: 'Give a man the secure possession of a bleak rock and he will turn it into a garden: give him a nine year's lease of a garden, and he will convert it into a desert.'

Nor was it only Smith who, 'in this as in many other cases', was systematically developing a naturalistic approach to social phenomena. He was in fact one of a small group, a main part of 'the Edinburgh Enlightenment'. This group also included, among others, that sometime chaplain to the Black Watch and later Edinburgh professor Adam Ferguson, the historian William Robertson, and – slightly older and starting to publish much earlier – David Hume.

It is to the point here to recall that Hume presented his own first shattering work not as an essay in conceptual analysis but as *A Treatise of Human Nature*; 'An attempt to introduce the experimental Method of Reasoning into Moral Subjects'. And, even where he is dealing with what is in the narrowest modern sense most strictly philosophical, Hume's characteristic contentions open the way to the open-minded discovery of causes which are altogether unlike and unsuggestive of their effects. In the first *Inquiry*, a work to which Darwin was attracted at the period of the M notebook, Hume contends: 'If we reason apriori anything may appear able to produce anything'; whereas the contrary assumption is 'the bane of all reasoning and free enquiry' (Hume, 1748, XII iii and IV i pp. 164 and 26).

(ii) The second point for us to take is that what Smith and the others were offering was evolutionary as opposed to creationist. The division of labour 'is not originally the effect of any human wisdom . . . It is the necessary consequence of a certain propensity in human nature'. Nine years earlier Adam Ferguson had made a similar point quite generally:

> Mankind in following the present sense of their minds, in striving to remove inconveniences, or to gain apparent and contiguous advantages, arrive at ends which even their imagination could not anticipate . . . Every step and every movement of the multitude, even in what are called enlightened ages, are made with equal blindness to the future; and nations stumble upon establishments, which are indeed the result of human action but not the execution of human design. (Ferguson, 1767, pp. 122–3)

The same seminal passage at once proceeds to enforce the point that – at any rate in default of sufficient independent evidence of their particular existence – there is no longer any call to postulate great creative culture heroes to explain the origin of such establishments:

> If we listen to the testimony of modern history, and to that of the most authentic parts of the ancient; if we attend to the practice of nations in every quarter of the world, and in every condition, whether that of the barbarian or the polished, we shall find very little reason to retract this assertion . . . We are therefore to receive, with caution, the traditionary histories of ancient legislators, and founders of states. Their names have long been celebrated; their supposed plans have been admired; and what were probably the consequences of an early situation is, in every instance, considered as an effect of design . . . If men, during ages of extensive reflection, and employed in the search of improvement, are wedded to their institutions, and, labouring under many inconveniences, cannot break loose from the trammels of custom; what shall we suppose their humour to have been in the times of Romulus and Lycurgus? (ibid.)

Durkheim once said in this connection, in his essay on 'Montesquieu and Rousseau, precursors of sociology', that the myth of the inspired and revolutionary legislator had, more than anything else, been the hindrance to the development of his subject. Notice, too, that there are parallel, indeed still more forceful objections to the hypothesizing of sudden creation not by an individual but by a collective.

Already in the *Treatise* Hume had deployed many of these objections in order to dispose of suggestions that the actual origins of all governments must have been in historical contracts. However, he allows, 'philosophers may, if they please, extend their reasoning to the suppos'd *state of nature*; provided that they allow it to be a mere philosophical fiction which never had, and never cou'd have any reality'. Later he speaks in precisely parallel terms about the legend of historical social contracts made to end that pre-governmental state of nature (Hume, 1739–40, III ii 2 and 8, p. 493).

Hume as a social scientist is concerned here with questions of how in fact institutions and constitutions originated. Was it through slow evolution or by sudden creation? So he is not going to ban all reference to some hypothetical or fictitious social contract from discussion of the very different questions of the legitimacy or otherwise of the actions of present-day governments – questions, that is, of justification.

His evolutionary and sociological insight is that not only government but also other fundamental social institutions neither in fact arose nor could have arisen through a contract from a pre-social state of nature; if only because promising itself already essentially presupposes the social institution of language.

Hume's own proposed outline solution to this problem of actual origins is subtle, hard headed, and profound; notwithstanding that some of the terms in which he states that solution must, unfortunately, suggest the sociologically unsophisticated crudities which he himself is striving to reject. Where his less enlightened opponents tell tales referring back to deliberate foresight and contractual agreement, Hume argues that the fundamental social institutions could not have originated from this sort of planning. What is possible is that recognitions of common interest will lead to the regulation of conduct in ways which are not, and often could not be, derived from prior contracts:

> Two men, who pull the oars of a boat, do it by an agreement or convention, tho' they have never given promises to each other. Nor is the rule concerning the stability of possession the less deriv'd from

human conventions, that it arises gradually, and acquired force by a slow progression . . . In like manner are languages gradually establish'd by human conventions without any promise. In like manner do gold and silver become the common measures of exchange . . . (ibid., III ii 2 p. 490)

To the philosopher, and not to the philosopher only, that penultimate illustration should be most impressive of all. To think that the natural languages, formations and richness and subtleties of which it is so hard even faithfully to delineate, must be in the main evolved by-products of the actions and interactions of people who were themselves, whether individually or collectively, incapable of designing anything of comparable sophistication (Austin). In what language, after all, would the select committee charged with the task of designing the first language have conducted its deliberations?

It is no wonder that Ferguson became lyrical about language:

This amazing fabric . . . which, when raised to its height, appears so much above what could be ascribed to any simultaneous effort of the most sublime and comprehensive abilities. . . . The speculative mind is apt to look back with amazement from the height it has gained; as a traveller might do, who, rising insensibly on the slope of a hill, should come to look from a precipice of almost unfathomable depth, to the summit of which he could scarcely believe himself to have ascended without supernatural aid. (Ferguson, 1792, I p. 43)

The eccentric and studiously old-fashioned Lord Monboddo, who surely knew as much about linguistics as any of his contemporaries, but who was not so fully seized of the evolutionary possibilities, takes up Ferguson's disclaiming hint of some supernatural aid. Monboddo 'can hardly believe but that in the first discovery of so artificial a method of communication, men had supernatural assistance'. So he is 'much inclined to listen to what Egyptians tell us of a God, as they call him, that is an intelligence superior to man, having first told them the use of language' (Burnett, IV p. 484).

(iii) Although Darwin was from the beginning interested in the problem of evolution of language he had in the event very little to contribute. In *The Origin of Species* he makes one reference to elaborate human languages – in the chapter on 'Mutual Affinities of Organic Beings: Morphology: Embryology: Rudimentary Organs' (Darwin, C. 1859, p. 406). He simply uses what was already known about the development of families of languages to make a point about the properly genealogical (as opposed to cladistic) classification of species. He had while working on his own problem been encouraged by the available findings of comparative linguistics. It was in 1786 that Sir William Jones had shown that a comparison between Sanskrit, Greek and Latin brought out similarities which could only be explained on the assumption that these three languages had originated from some ultimate common source, from what was perhaps now a dead language. This genealogical, evolutionary conception was extended in 1816 by Franz Bopp, who introduced the term 'Indoeuropean' to include Sanskrit, Persian, Greek (Ancient and Modern), Latin and most of the languages still spoken in Europe. In *The Descent of Man* Darwin had rather more to say about the necessary imitations of sounds than about the creation of concepts and the attachment of meanings.

Since at the human level behavioural variations can be and often are intended and intelligent innovations, these are presumably the materials upon which both natural and artificial selection work in generating linguistic evolution. So there should be scope here for the useful employment of another distinction arising in the field of the social sciences. It is, however, one which seems not to have been made either by the Scottish founding fathers themselves or by their most sympathetic modern interpreter, F. A. Hayek. In a posthumous masterpiece published a year or so later than *The Wealth of Nations*, a masterpiece with which Darwin was at one time surely familiar, Hume recognized that something which had not been designed either by one individual or by a committee might nevertheless be the ultimate product of innumerable more or less intelligent initiatives not consciously co-ordinated. In Part I of the *Dialogues concerning Natural Religion* Philo is scripted to say:

If we survey a ship, what an exalted idea we must form of the ingenuity of the carpenter, who framed so complicated, useful and beautiful a machine? And what surprise must we entertain, when we find him a stupid mechanic, who imitated others, and copied an art, which, through a long succession of ages, after multiplied trials, mistakes, corrections, deliberations, and controversies, had been gradually improving? (Hume, 1779)

The distinction needed is: between, on the one hand, social mechanisms producing results unintended by, and even contrary to the wishes of, those whose actions constitute the operation of these mechanisms; and, on the other hand, the generation of what may suggest brilliant individual or collective design through the not intentionally co-ordinated intelligent initiatives and responses of various persons or groups of persons, most of whom cannot have been directly acquainted with one another. It may be true that no great work of art was ever created in a series of committee meetings. But some of the greatest – *The Iliad* and *The Odyssey*, for instance – most probably were the ultimate achievement of successive generations, with many individual bards making their comparatively tiny contributions piecemeal and seriatim. And, whatever is the true answer to the Homeric question, we are entitled to be entirely confident that the original evolution of the first natural language, or the first natural languages, was – like the continuing present development of all the living languages of today – the half-blind and never intentionally co-ordinated work of innumerable relatively 'stupid mechanics'. Certainly it was never the spectacular once-for-all achievement of some great creative innovator, the linguistic Lycurgus or the philological Prometheus.

3. AN ATHEIST PROVIDENCE GUARANTEEING UTOPIA?

We move on now from questions about the influence and lack of influence on Darwin of Adam Smith and other great Scots of the Enlightenment to questions about the relations or lack of relations between Darwin and Marx. The first need is to put the record straight about two very different matters of considerable present interest.

(a) (i) The first of these is that Marx is supposed to have written to Darwin asking his permission to dedicate to him a further volume or a fresh edition of *Capital: A Critique of Political Economy*. This oft told tale – I confess to having told it many times myself – suggests that Marx was eager to put *Capital* on a par with *The Origin of Species*, as a second and comparable contribution to science. (Those of an older generation having literary interests may be reminded of George Bernard Shaw's denunciations of Shakespeare, denunciations presumably motivated by a parallel desire to have Shaw rated with Shakespeare in the top dramatic class.)

Whatever estimate Marx may have formed of *Capital*, whatever his aspirations for that book, this story has for several years now been known to be without foundation. The facts are as follows. Writing to Engels on 18 June 1862 Marx said: 'It is remarkable how Darwin recognizes among beasts and plants his English society with its division of labour, competition, opening up of new markets, "inventions", and the Malthusian "struggle for existence".' However, having so remarked, Marx makes nothing of it. On 16 June 1873, Marx sent an inscribed copy of the second German edition of *Das Kapital* to Darwin, who thanked him only on 1 October, apologizing for not understanding 'the deep and important subject of political economy'. This copy remains, its pages still uncut, in Darwin's library at Down House.

The Darwin letter which gave rise to this false story about the proposed dedication is dated 13 October 1880. It must have been addressed not to Marx but to his future common law son-in-law Edward Aveling, and have been referring not to *Capital* but to *The Student's Darwin*, published in 1881. The first suspicions were published by Lewis Feuer in *Annals of Science* for January 1975. These were confirmed when Thomas Carroll discovered the relevant Aveling letter, dated 12 October 1880, in the Cambridge University Library: this was published in the same journal in July 1976. The whole affair was sorted out in a definitive article by Lewis Feuer, 'The Case of the "Darwin–Marx" Letter', in *Encounter* for October 1978. One final twist in the tale is that it would probably never have got started at all had the leading Soviet Marx scholar, the former Menshevik Ryazanov, not fallen victim to the Great Terror.

(ii) A second miscellaneous matter for sorting out is the allegation of racism. Among the many articles sent to me by friends who knew I was working on the present book is one from *Spearhead* (no. 130, August 1979), treating Darwin as number 3 in a series 'Great British Racialists'. (This article, by the way, features the old dedication story; as do Steven Rose's 'Survival of the Fittest Philosophy' in *New Socialist* for July/August 1982 and Marcel Bloch's 1983 'Marxist Introduction' *Marxism and Anthropology*. However – credit where credit is due – it does not reappear in the centenary tribute 'Darwinism and Marxism' published in *Militant* for 16 April 1982). Fears or hopes similar to those of *Spearhead* may have been raised earlier by the subtitle of *The Origin of Species*: 'The Preservation of Favoured Races in the Struggle for Life'.

When talk turns to race and racism the first question which ought to be raised, and usually just about the last which is, is: 'What is it which it is proposed that we should mean by the word "racism"?' In particular, most crucially, we need to ask whether we are being urged to denounce and eschew those accounted racists in respect of their normative or in respect of their (would be) factual beliefs. If it is the former, if the racist is to be understood to be someone who wants to advantage or disadvantage individuals for no other or better reason than that they happen to be members of this racial group rather than that, then indeed the stance of the racist quite categorically ought to be repudiated utterly; and racism to be correspondingly execrated and abhorred. (I may perhaps be permitted to interject that some of my own more traumatic personal memories are of notices in National Socialist Germany proclaiming 'This village is Jew-free' or 'Jews not wanted here'.)

But, if, on the other hand, to be put down as a racist it is sufficient to believe that there are or may be hereditarily determined average differences in potentialities or in temperaments as between some racial groups and others, then forcibly to deny platforms to racists, to condemn and to refuse to finance or even to tolerate racist research, and so on, is radically illiberal and obscurantist. No one, surely, makes so bold as to deny that the usual defining characteristics of the main recognized human races are hereditarily determined; or to argue, hence, that the reason why there are so many black and brown skins to be seen in, say, Southall or Brixton must be the tropical intensity of the sunlight affecting those particular parts of London? So, if it is also allowed – as it surely must be? – that there is at least some hereditary element in the determination of some sorts of potentiality and temperament, then it must become probable that there will be some correlations between some of these and some definitionally racial characteristics (Flew, 1976a, I 5, 'The Jensen Uproar').

In the first interpretation of the word, making it refer to matters of normative belief, there seem to be absolutely no grounds for pillorying Darwin as a racist. On the contrary, for we have already seen that he fully shared his whole family's principled hatred for what was in contemporary practice always and only Negro slavery. It is worth quoting from the 3 July 1832 *Beagle Diary* entry. Speaking of the blacks, both slave and free, whom he had met in Brazil Darwin writes: 'I cannot help believing that they will ultimately be the rulers. I judge of it from their numbers, from their fine athletic figures, (especially contrasted with the Brazilians) . . . and from clearly seeing that their intellects have been much underrated; they are efficient workmen in all the necessary trades' (Darwin, 1832–6, p. 77; and compare Darwin, F., I p. 246).

Equally typical of a very different writer is the letter which Marx wrote to Engels on 7 August 1866. Marx had been reading, and been completely taken in by, P. Tremaux's *Origins, and Transformations of Man and Other Beings* (Paris, 1865). This Marx thought 'much more important and copious than Darwin . . . For example, the author . . . proves that the common Negro type is only a degeneration of a much higher one' To this Engels, to his credit, responded with two hatchet jobs on Tremaux. But this still did not inhibit either of them from regularly employing the word 'nigger' (in English) as a term of contemptuous abuse.

Again, in Darwin we can find references also to struggles for existence between human races. Thus in a very late letter, dated 3 July 1881, he wrote: 'The more civilized so-called Caucasian races have beaten the Turkish hollow in the struggle for existence. Looking to the world at no very distant date, what an endless number of the lower races will have been eliminated by the civilized races throughout the world' (Darwin, F., I p. 316).

If any observations on this topic, however neutral and detached in tone, are to contemporary 'liberal' tastes unacceptable, then how much more unacceptable those readers and writers ought to find the similar observations which abound in Marx and Engels – observations which are made there with approval, relish, even exultation. Take, for instance, an article which appeared over the signature of Engels, but certainly with the wholehearted approval of Marx, in the *Neue Rheinische Zeitung* for 13 January 1849: 'The general war which will then break out will smash this Slav Sonderbund and wipe out all these petty, hidebound nations down to their very names. The next world war will result in the disappearance from the face of the earth . . . of entire reactionary peoples. *And that, too, is a step forward*' (Italics supplied).

Anyone not too bigoted to recognize Marx as a lifelong anti-semite and, in the worst sense, a racist will profitably refer to Weyl. The 1844 pamphlet *On the Jewish Question* contains many statements which could have been quoted with approval, and for all I know were, either in *Der Stürmer*, the main and obscene National Socialist anti-semitic weekly, or in Ulrike Meinhof's dedicated vindication of the Holocaust, in her statement from the dock at the trial of the West German Red Army Fraction. The same pamphlet was by Lenin's Commissar for Education, Anatole Lunacharsky, commended as 'a true stroke of genius' and 'absolutely valid to this very day'. It remains a theoretical justification for the anti-semitic policies of the international socialists (as opposed to International Socialists) of the Soviet Union.

(b) Suppose now that we are prepared 'to put up with the crude English method of discourse' in Darwin. Presumably, although he chose not to expatiate, Marx was sighing here for Hegelian abstractions and a heavy teutonic appearance of profundity; and found it hard to bear with 'Cambridge . . . the English tradition of Locke and Berkeley and

Hume . . . a preference for what is matter of fact . . . prose writers, hoping to be understood' (Keynes, pp. v-vi). Whether we only bear with such Englishness, or positively welcome it, we have to ask how far Darwin's discoveries do provide any natural scientific support for the distinctive beliefs of Marx.

Certainly they do provide support for several beliefs which Marx held, and held strongly. Take, for instance the belief actually specified in that motto quotation from a letter to Lassalle: a colleague to whom elsewhere in the correspondence Marx refers variously as a 'water-polack Jew', 'Jew Braun', 'the little Kike', 'Izzy the Bounder' and 'Baron Izzy'; and over whose excruciatingly painful and protracted death Marx and Engels (in a letter of 22 August 1868) were later to gloat. No one persuaded that our species has evolved through natural selection in a struggle for existence should be surprised to find in human history a deal of conflict over the utilization of resources – over *lebensraum*, if you like, and *lebensmittel*. Yet we still ought to ask to hear argument before allowing to Marx that these conflicts are always or even predominantly class conflicts. (After all he himself, as we have just seen, sometimes so far forgets his official position as to allow that they are often fights to a finish between races or nations.)

Again, as we have also seen, Darwinism does offer massive support for materialism, construed as a doctrine of the primacy of matter over mind and consciousness. In particular it renders it almost impossible to present any even halfway plausible defence of a Platonic-Cartesian view of the nature of man; a view, that is, insisting that the real person is not the familiar creature of flesh and blood but instead an essentially incorporeal alien. But these beliefs, which Darwinism certainly supports, were and are by no means peculiar to Marx and his followers. Among Darwin's immediate contemporaries and associates, for instance, they were certainly shared by T. H. Huxley. Indeed, it was as an evolutionary biologist that T. H. Huxley like his grandson Julian, saw human multiplication as the supreme problem: 'The population question is the real riddle of the sphinx . . .'; and beside this 'all other riddles sink into insignificance' (Huxley, T. H., 1890, pp. 328–9; and compare pp. 315–16). By contrast Marx and Engels, except once in a very late letter by the latter, persistently refused to allow that there are or ever could be human population problems; a stance for which the Chinese and many other peoples are paying, and will I fear continue to pay, a monstrously heavy price (Flew, 1970, pp. 51–6; and compare pp. 40–3).

So, while Darwinism certainly reinforces some positions already strongly held by both Marx and Engels, we should get a very different answer when we press the question whether it supports what is peculiar and distinctive; whether, that is, it supports what in their correspondence with one another they distinguish as 'our view'. That view was first and most famously expressed in *The Communist Manifesto* of 1848. It can be characterized best and most briefly as apocalyptic historicist prophecy apparently based upon an all-embracing, supposedly scientific, philosophy of history. In this characterization the word 'historicist' is being employed in Popper's sense. A historicist is thus someone who proclaims inexorable supposed natural laws of historical development, and the specimen which Popper himself displays is taken from the Preface to *Capital*: 'When a society has discovered the natural law that determines its own movement, even then it can neither overleap the natural phases of its evolution, nor shuffle them out of the world by a stroke of the pen' (Marx, 1867, p. 10; Popper's translation).

The expression 'philosophy of history' is used here to refer not to the analysis of concepts employed by historians but to the attempt to discern patterns in the succession of events. In this thinnest understanding the first specimen of the species is perhaps the account given by Plato in Book VIII of *The Republic* of the supposedly natural and normal process of Greek constitutional degeneration – from the ideal city state ruled by Platonic philosopher kings to timocracy, from timocracy to oligarchy, from oligarchy to democracy, and from democracy to dictatorship. But it is only with the rise of Christianity that we begin to find philosophy of history encompassing all mankind. The story now is the story of the unfolding of God's plan for his special earthly creatures: Creation, the brief Age of Innocence, the

Fall, the Incarnation, the Second Coming, and the Kingdom of God on Earth. Secular developments are seen within this all-embracing divine framework. For instance: in the late third and early fourth century A.D. Eusebius of Caesarea composed a *Praeparatio Evangelica*, purporting to show that the history of the pre-Christian world should be regarded as a process designed to culminate in the Incarnation. Jewish religion, Greek philosophy, and the dominion of Rome constituted the seedbed in which the Christian revelation could be expressed and accepted: had the second person of the Trinity become incarnate at any very different time or in any very different place, then the world would not have been ready.

To any question why all these historical developments taken together 'should be regarded as a process designed to culminate in the Incarnation' Eusebius has, given the Christian Revelation, a simple and utterly decisive answer. The reason why it should be so regarded is because it was. To the further question, 'how can he be so sure about the consummation of the whole historical process in the Kingdom of God on Earth?' his answer, on the same Christian assumptions, can be equally simple and equally decisive. He can be sure because he has God's promise, and no power in all the universe can say Him nay: 'Who is he that saith, and it cometh to pass, when the Lord commandeth it not?' (Lamentations, 3: 37).

In *The Communist Manifesto* Marx and Engels expound a philosophy of history which is at several points reminiscent of the Eusebian scheme. It is indeed, as Soviet spokespersons so love to say, no accident that it is. For Marx and Engels were, or had been, Young Hegelians. And Hegel's philosophy of history was a secularization of what had traditionally been accepted among Christians. When Marx and Engels, as they put it, 'stood Hegel on his head or, rather, stood him right way up' they may have replaced an idealist by a materialist. Yet they still had a Hegel.

Consider, for instance, how the promised eventual establishment of the Kingdom of God on earth is replaced by the promised coming total triumph of the class to end all classes; and how, at least in later editions, the Fall in the Garden of Eden makes way for the unexplained lapse from primitive communism into a property-owning, and hence, supposedly, class-divided society. Notice too how Marx and Engels – like Alfred Hitchcock in a Hitchcock movie – find a place in the plot for themselves: '. . . a small section of the ruling class cuts itself adrift, and joins the revolutionary class, the class that holds the future in its hands' (Marx and Engels, 1848, p. 91). But here the crucial message of redemption is brought not by one element of the triune Godhead but by the proletarians' PhD: '. . . a portion of the bourgeoisie goes over to the proletariat, and in particular, a portion of the bourgeois ideologists, who have raised themselves to the level of comprehending theoretically the historical movement as a whole' (ibid., p. 91).

Suppose now that we ask Marx and Engels the questions put previously to Eusebius of Caesarea. How do they justify the initial and later amended claim: 'The history of all hitherto existing society is the history of class struggles'? (ibid., p. 79) What, above all, is their warrant for the apocalyptic eschatology: 'What the bourgeoisie . . . produces . . . is its own grave-diggers. Its fall and the victory of the proletariat are equally inevitable'?; and so 'In place of the old bourgeois society, with its classes and class antagonisms, we shall have an association, in which the free development of each is the condition for the free development of all'? (ibid., pp. 94 and 105)

It is, of course, this bold and enormously confident historicist prophecy, reiterated in so many later works, which has been and continues to be the main source of the appeal of 'our view'. In the nineteenth century a very typical anonymous Russian convert from populism exclaimed: 'The knowledge that we feeble individuals were backed by a mighty historical process filled one with ecstasy and established such a firm foundation for the individual's activities that, it seemed, all the hardships of the struggle could be overcome' (quoted Wesson, p. 46). In our own time we have heard Nikita Khrushchev boast: 'Communism is at the end of all the roads in the world. We shall bury you.'

In fact, the confidence placed by Marx himself in such historicist contentions about the allegedly inexorable destinies of bourgeoisie and proletariat appears originally to have been founded upon nothing more concrete than a piece of high Hegelian philosophical analysis: 'Marx', as Lesnek Kolakowski says at the start of his great three-decker *Main Currents of Marxism*, 'is a German philosopher' (Kolakowski, I p. 1). I will here spare the reader any specimens of such 'philosophical analysis', having provided enough elsewhere (Flew, 1984, § 3). What must be said is that in *The Communist Manifesto* Marx and Engels developed a sketchy sociological argument, from unevidenced and dubious premises: 'The advance of industry, whose involuntary promoter is the bourgeoisie, replaces the isolation of the labourers, due to competition, by their revolutionary combination, due to association. The development of Modern Industry, therefore, cuts from under its feet the very foundations on which the bourgeoisie produces and appropriates products' (Marx and Engels, 1848, pp. 93–4).

For the rest of his life the most important academic work of Marx, work from which he continually allowed himself to be distracted and diverted, consisted in attempts to build solid empirical foundations for these historicist contentions, contentions so essential to 'our view'. Marx signed his first contract to produce the book building these necessary foundations in 1845. Although he accepted, and retained, an advance of 1500 francs no manuscript was ever delivered. Nor was he in the end satisfied with *Capital*; which, significantly, he never finished. (The best source for the whole story is Schwarzschild.) Nevertheless, whether with or without its necessary empirical foundations, 'our view' was aggressively and successfully marketed as 'scientific socialism'. *Socialism: Utopian and Scientific* by Engels was throughout the nineteenth century, and perhaps is still, the most studied work in the whole Marx–Engels literature.

To understand the intended contrast notice a passage found among the Marx papers only after his death. It was originally intended to go into *The Civil War in France* (1871), but has been crossed out. Referring to Fourier, Cabet, Owen, and other utopian socialists seen off at the end of *The Communist Manifesto* he writes:

They attempted to compensate for the missing historical preconditions of the movement with fantastic pictures and plans of a new society . . . From that moment on, when the working-class movement became a reality, the fantastic utopians disappeared: not because the working class abandoned the objective for which these utopians had reached, but because the true means had been found for its realization . . . Still, both the ultimate aims announced by the utopians (i.e. the end of the system of wage-labour and class domination) are also the ultimate aims of the Paris Revolution and the International. Only the means are different. (quoted Lasky, pp. 38–9)

At last we are in a position to give a short, sharp answer to the question from which sub-section 3(b) began: 'Does Darwinism provide natural scientific support for "our view"?' No, it does not! It is one thing, and surely not unreasonable, if you believe that we are creatures designed by Omnipotence in order to fulfil his purposes, and if you also believe that the establishment of the Kingdom of God is a part of those purposes, to put your faith in the ultimate fulfilment of this eschatological promise. It is quite another thing, and laughably incongruous, if you have rejected God and accepted a Darwinian account of the origin of species, to insist that a new kind of secular and scientific providence is bound to have its way; that the annihilating victory of a desperate and ever more impoverished proletariat is absolutely guaranteed to produce, in fairly short order, a conflict-free utopia. For there can be no doubt but that, for instance, the second clause of the concluding sentence of the main body of *The Communist Manifesto* is intended to assert, not a tricky tautology, but a substantial revelation: 'In the place of the old . . . society . . . we shall have an association in which the free development of each is the condition for the free development of all' (Marx and Engels, 1848, p. 105).

(c) But now, even though we have had to conclude that Darwin's discoveries provided no support whether direct or indirect for the distinctive position of Marx and Engels, perhaps it is still possible to make out the claim made by Engels in his address at the graveside? Perhaps it does have, nevertheless, to be conceded that Marx and Marxism are, scientifically speaking, on all fours with Darwin and Darwinism? Certainly this is a claim which – with or without the feeble prop of the false dedication story – continues to be made. It has been made by almost every professional scientist either raised as a Marxist or converted to Marxism, including both the anthropologist and the biologist mentioned in 3(a)(ii), immediately above. Lenin himself hath said it. For on the 1894 pamphlet 'What the "Friends of the People" are' he wrote, with reference to *Capital*: 'It will now be clear that the comparison with Darwin is perfectly accurate' (Lenin, p. 21). And in 1968 the official philosopher to the Communist Party of Great Britain (Muscovite) deployed an even bolder because more specific contention: 'The methodology by which Marx arrived at his theory of social development is exactly the same as that employed by Darwin in establishing the theory of the evolution of species by natural selection' (Cornforth, p. 27).

The general question whether Marxism constitutes a corpus of knowledge and a tradition of inquiry meriting the diploma title 'science' cannot be contained within the scope of the present book. But dealing with the more specific contentions provides useful occasion to bring out points about Darwin by contrasting his methods with those of Marx. This comparison will surely show that it was because Engels was so right in his insistence in the second part of his obituary speech that his friend had been before all else a revolutionary that he was so wrong to conclude the first part with the claim: '*So war dieser Mann der Wissenschaft*' (That was this man of science).

(i) The first point concerns historicism. The abundant biographical evidence makes it totally beyond legitimate dispute that Marx came to and pursued his researches in economics and sociology in order to find empirical and scientific warrant for the historicist prophecies made most famously, if not strictly first, in the *Manifesto*. These prophecies, as was said earlier, were by Marx and Engels originally founded upon the most arbitrarily *a priori* Hegelianism: the two leaders had, like Moses Hess, been of the few who had 'taken the philosophical road to Communism'. But if revolutionary troops – the poor, bloody infantry of the class struggle – were to be assembled and encouraged by the assurances of inevitable victory and inevitable utopia, then those assurances had to be presented as scientifically based. Anyone who has read H. A. L. Fisher's *Our New*

Religion will appreciate that Marx and Engels had the same compelling reason to offer their revelation as *scientific* socialism as Mrs Mary Baker Eddy had for offering hers as Christian *Science*. So in the *Communist Manifesto* – a document in almost every other way at an opposite extreme to *Science and Health* – Marx and Engels were, as it were, writing a monster post-dated cheque drawn on an account in the Bank of Scientific Knowledge, an account into which they had not so far been able to deposit any funds.

Now no one should make the mistake of dismissing any suggested hypothesis as unscientific simply on the grounds either that its antecedents are in some way disreputable or that those putting it forward were persuaded of its truth even before any proper testing had begun. It is not only great mathematicians who persist in following with unshakeable conviction hunches which they eventually prove to have been correct. Kepler himself began his calculations believing with his whole heart and mind that the planetary orbits must be circular, while his own grounds for so believing were entirely metaphysical and non-observational. In the end, though the end was long delayed, he was persuaded, or persuaded himself, to try ellipses as the next closest and hence in his view the next best thing: which, by looking over his shoulder, we know to have been the right answer.

Neither of these two grounds would by itself warrant refusal to permit Marx to intrude on to Darwin's pedestal. The first objection is, neither that the main economic-historical hypotheses of Marx originated in disreputable philosophy, nor that he was most stubbornly committed to them on the basis of evidence which at least at first it would be an understatement to describe as insufficient. These are, certainly, grounds for unease. But, for the reasons given, they cannot be alone decisive. The first decisive point is that those hypotheses which he himself saw as most crucial and most distinctive are historicist; and that every historicist hypothesis could have been known to be false already, previous to any further and more particular investigation.

In a letter to Weydemeyer dated 5 March 1852 Marx tells us very clearly what he saw to be most crucial and most distinctive in his own supposed scientific achievement, although he is too discreet to unveil any secrets of the sources of his supposed knowledge:

> What I did that was new was to prove: (1) that the *existence of classes* is only bound up with *particular, historic phases in the development of production*; (2) that the class struggle necessarily leads to the *dictatorship of the proletariat*; and (3) that this dictatorship only constitutes the transition to the *abolition of all classes* and to a classless society.

The second two assertions are, it is manifest, historicist. They assert, that is to say, laws of historical development stating that certain things are going to happen necessarily and unavoidably and irrespective of anything and everything which anyone does or might do in hopes of preventing these outcomes.

Against historicism in this understanding there is an objection more fundamental and more elementary than anything urged in Popper's own famous critique (Popper, 1957; and compare Flew, 1983a). Because we are agents who as such could do otherwise than we do do, there neither are nor can be any natural laws determining human action: there is, in other words, a flagrant conceptual incompatibility between the notions of action and of the impossibility of all alternatives. Furthermore, it can and must be suggested, both these two opposite and incompatible notions both are and could only be given in our most familiar experience of being able to do some things and not others. How, that is to say, could any of us even understand the relevant terms without being directly acquainted with specimens of both the two categorically different alternatives which those two (sorts of) terms (and expressions) are employed to pick out? It is a moment to reflect again upon the passage quoted from Locke in 1(b)(ii), immediately above (Flew, 1978, III and VII; Flew, 1982b; and Flew, 1983a).

It would, no doubt, be over-ambitious to maintain that Darwin saw all the implications of this human peculiarity of choice. But it would be far more wrong not to insist that, whereas Darwin in drawing a lesson from Malthus did appreciate that choice makes a vital difference, Marx seems never even to have begun explicitly to question the possibility of finding natural laws determining human action. And, if these are indeed a will-o'-the-wisp, then how much more implausible are his own pretended revelations of laws of historical development?

(ii) That first objection was a matter of error pure and simple. The rest are all more scandalous and less venial. They raise profound and significant questions about good faith. It should be obvious that anyone who with any pretensions to rationality is sincerely pursuing any objective, whether that objective is truth about some subject of theoretical inquiry or the attainment of some objectives of practical policy, will be concerned to monitor their success or failure in these pursuits; and equally concerned to adapt their tactics to the lessons of that success or failure (Chapter I 4(a), above). Such sincere and rational concern is also bound to manifest itself in lucid and unambiguous expression. For how, where there is obscurity or ambiguity, can we hope to settle the questions whether assertions were true or policies effective? Among the best *Maxims* of the Marquis de Vauvenargues are: 'Obscurity is the kingdom of error'; and 'For the philosopher clarity is a matter of good faith'. But not for the philosopher only.

It is in the harsh light of these fundamental and, once well and truly stated, incontestable principles that we have to interpret the often deplored fact that nowhere in all the enormous bulk of the collected works of Marx and Engels is there any systematic, unambiguous and tolerably full statement of what 'our view' was. Here as in other places the contrast with Darwin is, of course, absolute. Years before he ventured to publish anything on evolution by natural selection Darwin had for his private purposes written that 'sketch of my species theory', which he then took immediate steps to ensure should appear forthwith after his death were he to die before he was ready for publication. The businesslike, unambiguous, unevasive lucidity which characterized both Darwin personally and all his writings has often been commended; though it was to draw from Marx the reproach that Darwin was, as he was proud to be – to quote the speech of Elizabeth the Great at Tilbury – 'mere English'.

Deeply damaging passages in the Marx–Engels correspondence show that Marx took the points just now put by the Marquis de Vauvenargues, but that Marx was by no means consistently concerned that there should be thorough monitoring either of the propositions which he asserted or of the policies which he laboured to promote. (In view of the amount of dirt which is still to be found in that correspondence, it is mind-boggling to try to imagine how much worse were the bits which Engels confessed to having systematically destroyed after his friend's death.)

On the Vauvenargues point a passage from a letter dated 15 August 1857 is especially notable, the more so as it reveals much if not all of what Marx meant by the dialectics (or the dialectic method) which are (or is) in the published works, though never explained, sometimes commended:

> I took the risk of prognosticating in this way, as I was compelled to substitute for you as correspondent at the *Tribune* . . . It is possible I may be discredited. But in that case it will still be possible to pull through with the help of a bit of dialectics. It goes without saying that I phrased my forecasts in such a way that I would prove to be right also in the opposite case.

Can we, should we, refrain from repeating that tribute from Engels: '*So war dieser Mann der Wissenschaft*'?

(iii) Obscurity or ambiguity of statement, and subsequent further obfuscation, are in the present context important mainly as means of avoiding the admission that a statement has been falsified (shown to be false) or a policy discredited (shown not to produce the goods promised). We have already seen in Chapters I and II how constantly scrupulous Darwin was to consider and to try to meet every objection. When, as was the case with Lord Kelvin's estimates of the youth of the earth, there was at the time no satisfactory answer available; then Darwin – once he had satisfied himself, both that and what future discoveries might save the situation, and that his was by far the most plausible theory anywhere in the running – honestly recognized and recorded the difficulty, and left it to the future to meet. What more or what else could we ask of any scientist? Darwin, indeed, is a very paradigm of complete good faith, and single-mindedness in the pursuit of truth. It was altogether characteristic of both men that Freud could commend 'the great Darwin' because 'he made a golden rule for himself, writing down with particular care observations which seemed unfavourable to his theory, having become convinced that just these would be inclined to slip out of recollection' (Freud, 1922, p. 61).

With Marx, I fear, we have a very different story and a very different man. Two examples of his ways of dealing with falsifying fact will have to suffice. But both the two chosen are of the highest intrinsic interest and importance. Both help to show that Marx was, as Engels also said, first and always a revolutionary; albeit, as he neither said nor perhaps ever noticed, a revolutionary for revolution's sake rather than for the sake of whatever reliefs to man's estate his revolutions might rationally be expected to bring. It was, surely, this consuming revolutionary enthusiam which led him to commit these and other malpractices with falsifying fact?

First, then, consider what is customarily called the 'immiseration thesis'. In *The Communist Manifesto* this is asserted in what would appear to be a strictly self-contradictory form: 'The modern labourer, instead of rising with the progress of industry, sinks deeper and deeper below the conditions of existence of his own class' (Marx and Engels, 1848, p. 93). Anyone who has lived through successive British national incomes and prices policies will recall nothing so much as that favourite trades-union spokesperson's cry, that, while it stands to reason that everyone should get at least the average increase, some groups and, in particular, – surprise, surprise! – that represented by the speaker must get a lot more than the average.

Eschewing on this occasion, however, all such logic-chopping around that particular *Manifesto* formulation, we need only to understand that and why Marx and Engels need to maintain a strong immiseration thesis. For their revolutionary purposes they have to insist that the proletariat must inevitably grow progressively more numerous and poorer, while the capitalists must equally inevitably and at the same time grow fewer and richer. They have to, in order to sustain their cherished historicist prophecy: 'What the bourgeoisie, therefore, produces, above all, is its own grave-diggers. Its fall and the victory of the proletariat are equally inevitable' (ibid., p. 94).

For this transcendent revolutionary purpose there is no escape into any of those weakening, even if in the end still false, 'interpretations' so beloved of devout defenders never willing to allow their chosen and cherished gurus to be just plumb wrong. It is, thus, no use Marx and Engels saying that – actually – they meant relative not absolute immiseration, or that it is only one or other kind of tendency and not historicist law (1(b) ii, immediately above), and so on. They have either to pack it in or to brazen it out. Precisely and only because they were, first and always, revolutionaries rather than scientists they chose the second alternative. Thus in the first edition of *Capital* various available British statistics are given up to 1865 or 1866, but those for the movement of wages stop at 1850. In the second edition all the other runs are brought up to date, but that of wage movements still stops at 1850 (Wolfe, p. 323).

A stinging comment on this sort of thing was provided by Marx himself, in the course of a discussion comparing Malthus unfavourably and, I consider, unfairly with Ricardo. This discussion occurs in a manuscript published only posthumously. The key passages are most easily to be found in Karl Wittfogel's classic treatment of the topic of our second illustration of the differences between the methodologies of Marx and Darwin, *Oriental Despotism*. A scholar, Marx was on that occasion insisting, should seek the truth in accordance with the internal needs and logic of science, no matter how this might affect any class or other interest; and he praised Ricardo for taking this line, which Marx called 'not only scientifically honest but scientifically required'. By contrast, 'a man who tries to accommodate science to a standpoint which is not derived from its own interest, however erroneous, but from outside, alien, extraneous interest, I call mean [*gemein*].' Marx was, therefore, wholly consistent when on the same occasion, and altogether against the spirit of Leninist partisanship [*partinost*], he paid tribute to a class enemy: 'As far as this can be done without sin against his science, Ricardo is always a philanthropist, as he indeed was in practice' (Wittfogel, pp. 386–7).

It was only after taking up residence in London and beginning a really thorough study of the classical economists that Marx was led to recognize 'the Asiatic mode of production' as a form of political economy radically different from anything for which he and Engels had previously provided. Adam Smith had noted similarities between imperial China and 'several other governments of Asia'. James Mill in his *History of British India* identified the 'Asiatic model of government' as a general institutional type, and he rejected attempts to assimilate it to feudalism. His son John Stuart Mill accepts this identification right from the earliest pages of his *Principles of Political Economy*, a work first published in 1848, the same year as the *Manifesto*. We know that between 1850 and the summer of 1853 Marx studied both *The Wealth of Nations* and these two works by the Mills; as well as Prescott's *Conquest of Mexico* and *Conquest of Peru* and – what had been a prime source for the economists – François Bernier's *Travels in the Mogul Empire: A.D. 1656–1668.*

Having in and through all this reading seized on the notion of 'Asiatic society' (Richard Jones) or 'Oriental society' (J. S. Mill) as a quite distinctive social form, Marx employed it in his articles for the *New York Daily Tribune*, as well as in early drafts for *Capital* and in *Capital* itself. He seems indeed never to have abandoned the idea. But – in part no doubt because, as we have remarked, there was at no time any precise, full, yet compassable statement of 'our view' – he never begins to face the question of how if at all that theoretical scheme can be squared with or adjusted to this admission of this further kind of social system.

He ought to have seen this as a problem right from the beginning. For, even though attention was concentrated on the final transformation in the series, *The Communist Manifesto* had provided for a serial evolution of species of social system: ancient society developing into feudalism, the bourgeois or capitalist order emerging from the womb of feudalism, and then the final stage in which the class to end all classes shatters capitalism and establishes the classless socialist utopia. To fit in a primal stage of primitive communism is not too difficult. But to attempt to intrude a fifth kind of social order within the original four-part serial might be thought to be rather like trying to squeeze other inhabited worlds into a scheme which had already embraced a unique incarnation at the very centre of the physical universe and its history. Did Marx never explore what would seem the obvious escape route, that of arguing that the Roman Empire under Septimus Severus or Domitian had itself become a specimen of the genus 'Oriental despotism'?

Yet even if that route would go, there are other difficulties of which he cannot dispose so easily. The opening words of *The Communist Manifesto* had been: 'The history of all hitherto existing society is the history of class struggles.' It was easy to amend that by adding the qualification 'written', thus making room for an Age of (propertyless) Innocence prior to the (appropriative) Fall. It is a very different thing to meet the challenge to find classes and class struggles in the indisputably post-lapsarian 'Asiatic system'. It is very different and, for Marx, very difficult because this challenge is at one and the same time a challenge both to his theoretical and to his practical commitments. The way in which he in fact met or, rather, failed to meet it constitutes both a 'sin against his science' and a sinister indication that, even on the most charitable judgement, neither Marx nor Engels can be allowed to have been unambiguously and unreservedly committed to libertarian and humanitarian ends.

It is, I suggest, manifest how the sociological principles of Marx required him to respond to this theoretical challenge: he 'should have designated the functional bureaucracy as the ruling class of Oriental despotism' (Wittfogel, p. 381). But, even had he taken this self-appointed road he would quickly have found himself in what would have been, from his point of view, a cul-de-sac. For under Oriental despotism this ruling class is so dominant that it can and does prevent all organization by any possible rivals: 'The history of hydraulic society suggests that class struggle, far from being a chronic disease of all mankind, is the luxury of multicentred and open societies' (ibid., p. 329).

Even granting this, you may ask, why should Marx not have met all such objections by simply conceding that his claims about class conflict applied only to those Western societies in which he and his readers were most interested? The 64,000-dollar answer is that the centralized, totally socialist state proclaimed in *The Communist Manifesto* – that supposedly quite new form of social order for which Marx was working all his life – would be sure to be, or very soon become, a kind of Oriental despotism; and, so far from the state machinery of such a social system tending to wither away, historic experience shows Oriental despotism to be an extraordinarily persistent form, with no strong inherent tendencies either towards decay and dissolution or towards development into some different and perhaps more pluralist system (ibid., passim).

There is no call and no place here for us to deploy much argument in support of these conclusions. For the ambivalent and evasive attitudes of Lenin himself, the official suppression of all discussion of the concept of Oriental despotism in the now numerous established Marxist-Leninist societies, combine to show that both the 'New Tsars' and the 'New Mandarins' realize how well it applies. They not only see but also insist that socialism is in practice incompatible with individualistic dissent and democratic pluralism. Consider, for instance, the statement issued in 1971 by the Institute of Marxism-Leninism in Moscow. With its eyes then mainly on Chile and France it sketched a programme for achieving, through 'United Front' or 'Broad Left' tactics, irreversible Communist domination: 'Having once acquired political power, the working class implements the liquidation of the private ownership of the means of production . . . As a result, under socialism, there remains no ground for the existence of any opposition parties counterbalancing the Communist Party' (Quoted Flew, 1976a, p. 128).

Marx, of course, did not have our century's experience of Marxist socialism in practice. But he did have access to information about despotic systems under which the economy was largely or wholly socialist. (Noting that he read Prescott during his studies of 'the Asiatic mode of production', one is reminded that the socialist generals who ruled Peru for much of the 1970s called their project 'The Inca Plan'. Compare also Shafarevich, Part II.) Even more significant is the fact that Marx was again and again confronted with the charge that socialism, or at any rate socialism as he and Engels understood it, must involve slavery and despotism. Ruge, for instance, with whom Marx collaborated in 1844 on the *German-French Yearbook*, called the socialist dream 'a police and slave state'. In 1848 the vice-president of Louis Blanc's party told Engels: 'You are leaning towards despotism.' Similar objections were developed later by Proudhon and by Bakunin – by the latter in an 1873 book *Statism and Anarchism*.

To such criticism neither Marx nor Engels ever published any reply, although Marx did eventually add extensive annotations to his copy of *Statism and Anarchism*. Since both were as much inclined to engage in polemic as Darwin was reluctant, how can this silence, sustained over thirty years, be explained otherwise than as an indication that neither was able to think up any effective answer; but that, although they did not want potential followers to become persuaded of the correctness of these objections, they themselves were not seriously disturbed by them, or even, perhaps, positively pleased? The unlovely practical commitments of Marx thus seduce him into sins against science: not only, as we have seen, into his inhibited treatment of traditional Oriental despotism; but also into his refusal to apply what was there to be learnt to reveal the prospect of Marxist socialism realized as a prospect of Oriental despotism redivivus. The upshot, therefore, of this whole unconscionably long Section 3 is to display the hopelessness, not to say the sheer indecency, of any project to put Marx forward as the equal of Darwin (Shafarevich, VII 2).

4. THE CHALLENGE OF SOCIOBIOLOGY

In 1975 the Harvard biologist Edward O. Wilson, already well known for his work *The Insect Societies*, published *Sociobiology: The New Synthesis*. This is a well-illustrated, massive, monumentally learned book of nearly 700 pages. It was widely and favourably reviewed in the scientific, literary and lay press. But it also occasioned an explosive outburst of fury and execration from various protestors. Such opponents, although most of them are employed as scientists, were typically concerned not so much with whether what Wilson had actually said is in fact true as with its putative political and social implications: whether these implications were implications which they themselves thought they saw; or whether they were only implications which they believed that others had seen, or might in the future be misled to see. Despite their scientific professions these protestors appear to have only a slight and marginal interest in either what the biological truth is or in what, if any, practical morals could be validly derived therefrom.

Thus, there is in *The New York Review of Books* for 13 November 1975, a statement 'Against "Sociobiology"' by the Sociobiology Study Group of Science for the People – presumably, as usual, only the right left people. This says: in general, that 'Wilson joins the long parade of biological determinists whose work has served to buttress the institutions of their society by exonerating them from responsibility for social problems'; and, more particularly, that he has been trying to reinvigorate theories that once 'provided an important basis for the enactment of sterilization laws and restrictive immigration laws by the United States . . . and also for the eugenics policies which led to the establishment of gas chambers in Nazi Germany'.

(a) Suppose that we turn to Wilson's text. We find that the brouhaha has been occasioned by only two of his twenty-seven chapters: the first on 'The Morality of the Gene'; and the last on 'Man: From Sociobiology to Sociology'. Chapter 1 is exceedingly short. But it is long enough to contain a definition of 'sociobiology' as 'the systematic study of the biological basis of all behaviour'; to which the author adds the gloss that this book attempts 'to codify sociobiology into a branch of evolutionary biology' (Wilson, 1975, p. 4).

One anticipatory anxiety arises from the statement that 'the hypothalamus and limbic system . . . evolved by natural selection. That simple biological statement must be pursued to explain ethics and ethical philosophers, if not epistemology and epistemologists, at all depths' (ibid., p. 3). Since every explanation is an answer to a question, the temptation is to take this statement as a licence to insist that evolutionary biology can and must provide the answers to every possible question. But this would be both mistaken and dangerous.

Certainly evolutionary biology may have something if not necessarily everything to say about the answers to questions about how as a matter of fact we have come both to hold whatever would-be factual beliefs we do hold, and to support whatever norms we do in fact support. But the questions whether these would-be factual beliefs are true, and whether our evidence for holding them is sufficient to justify a knowledge claim, can be biological questions only when the beliefs themselves are biological. The questions whether and in what way it is rational to support the norms which we do support are again not biological questions; while the more particular question of how far, if at all, evolutionary biology is here even relevant must be deferred to a final, brief Chapter IV.

The great danger of failing to make the fundamental distinctions between sorts of questions is that we may conclude, falsely, that any which cannot be answered by evolutionary biology cannot be put and answered at all. We shall then find ourselves in the intellectually suicidal company of contemporary ultras of the sociology of belief. These people have become committed to maintaining that (thanks to the insights provided by their discipline they know that) there is no such thing as knowledge; no such thing, that is, as evidentially warranted true belief. They can admit only perceived or believed knowledge; but never the real thing, without prefix or suffix knowledge (Flew, 1983b). In the case of norms we shall, as we will see later, find ourselves with apparently no alternative but to accept as authoritative the mandates of pre human evolution.

(b) Turning to the far fuller Chapter 27, what calls most for remark is what it does not contain.

(i) Anyone approaching by way of the first three sections of the present chapter should be astonished that Wilson does not begin by insisting upon the supreme significance for human sociobiology of the inexpugnable and surely at least in very large part distinctively human reality of choice. In relating the conceptual schemes of Malthus and of Darwin in Section 1, immediately above, the main task was to bring out how no theory in the natural sciences needs to, but every theory in the human must, take account of this reality. Recognition of this reveals drastic limitations on the possibilities of genetic explanation. Since we already know that precious little if any of our significant social behaviour is, strictly speaking, subject to instinctual necessitation, we are already also in a position to know that the geneticist can hope only, albeit most importantly, to tell us that and why some of us cannot while others can learn to behave in certain sorts of ways, or that and why some or all of us are very strongly inclined or disinclined to take this course or that.

Of course, Wilson never denies any of this. But what he fails to do is to make it, what it surely should be, his starting point. Instead he cites findings by his Harvard colleague Richard Lewontin to show that there is not much genetic diversity as between different human populations. Having cited these findings Wilson quotes a powerful yet still rationally restrained environmentalist manifesto: 'Culture is not inherited through genes, it is acquired by learning from other human beings . . . In a sense, human genes have surrendered their primacy in human evolution to an entirely new nonbiological or superorganic agent, culture. However, it should not be forgotten that this agent is entirely dependent on the human genotype' (Dobzhansky, 1963).

On this Wilson comments: 'Although the genes have given away most of their sovereignty, they maintain a certain amount in at least the behavioural qualities that underlie variations between cultures.' Since moderately high hereditability has been shown in various dispositional characteristics, 'Even a small portion of this variance invested in population differences might predispose societies towards cultural differences' (Wilson, 1975, p. 550). Indeed it might. Nevertheless, this said, we have at once to underline the key words 'dispositional' and 'predispose'. For such genetic determination of individual dispositions necessitates, at most, the having of that disposition, not its behavioural actualization. We discover here in dispositions another token of the favourite Leibnizian type of cause, inclining yet not necessitating.

Wilson's own chief example is a disposition towards homosexual sex: '. . . the influence of genetic factors toward the assumption of certain *broad* roles cannot be discounted.

Consider male homosexuality . . . Kallmann's twin data indicate the probable existence of a genetic predisposition toward the condition' (ibid., p. 555). Another much bruited suggestion *not* urged in Chapter 27 is that the almost but not quite universal taboo on incest may have some genetic basis; a suggestion which, once the limitations on the possibilities of genetic explanation in the present context have been understood, cannot be dismissed simply by referring to incest effectively required in the Egyptian royal family or defiantly committed in Swedish films.

(ii) The second remarkable negative observation is how little the actual book *Sociobiology* contains which can do more than begin to explain the subsequent brouhaha. There seem to be none of those words or phrases which, for better or for worse, always touch off the Radical reflexes: nothing which such people could be expected to find half so offensive as, say, the subtitle of *The Origin of Species* – 'The Preservation of Favoured Races in the Struggle for Life'. And one might have thought that they would have been mildly pleased by the reference to 'the planned society – the creation of which seems inevitable in the coming century', even though Wilson does go on to explain how 'In this, the ultimate genetic sense, social control would rob man of his humanity' (ibid., p. 575).

No doubt some hyper-sensitive hackles were raised by the surely unexceptionably true statement: 'Human beings are absurdly easy to indoctrinate – they *seek* it.' Or perhaps those who think of Social Darwinism as always and only a 'right-wing' doctrine were upset by Wilson's urging of the possibility that 'the time has come for ethics to be removed temporarily from the hands of the philosophers and biologicized' (ibid., p. 562). Enough, however, of such aetiological speculation. Suffice it to say that the *Sociobiology* uproar has, if nothing else, demonstrated that the present Radical generation is much narrower and much more intolerant than its predecessors of the 1930s and '40s. How many, for instance, of those who abominated H. J. Eysenck's book *The Inequality of Man* realized that its title was mischievously borrowed from that of an excellent volume of essays by J. B. S. Haldane, for most of that earlier period both Britain's leading geneticist and a member of the Communist Party?

(c) What Wilson calls 'the central theoretical problem of sociobiology' emerges only in Chapter 1 and is not taken up again in Chapter 27. This problem is 'how can altruism, which by definition reduces personal fitness, possibly evolve by natural selection?' The solution is 'kinship: if the genes causing the altruism are shared by two organisms because of common descent, and if the altruistic act by one organism increases the joint contribution of these genes to the next generation, the propensity to altruism will spread through the gene pool' (ibid., pp. 3–4). There is a happy and characteristic tale of how back in the 1920s or '30s J. B. S. Haldane on some social occasion did various calculations to this effect, literally on the back of an envelope (Gould, 1978, p. 262).

Sympathetically understood, this is all as elegant as it is conclusive. It provides as effective an illustration as we could wish of how, in Neo-Darwinian theory, 'Natural selection is the process whereby certain genes gain representation in the following generations superior to that of other genes located at the same chromosome positions' (ibid., p. 3). Unfortunately, these sound and quite unanthropomorphic notions have since been taken up and expanded into a major exercise in popular mystification, Richard Dawkins's *The Selfish Gene*.

This is a work of popularization fully as bad in its own ways as either *The Naked Ape* or *The Human Zoo*. Whereas those works by Desmond Morris offer as the results of zoological illumination what amounts to a systematic denial of all that is most peculiar to our species contemplated as a biological phenomenon, Dawkins labours to discount or depreciate the main upshot of fifty or more years' work in genetics – the discovery that the observable traits of organisms are for the most part conditioned by the interactions of many genes, while most genes have manifold effects on many such traits. For Dawkins the main means for producing the obfuscatory specialities of the house is to attribute to genes characteristics which can significantly be attributed only to persons. Then, after insisting that we are all the choiceless creatures of our genes, he infers that we cannot but share the unlovely personal characteristics of those all-controlling monads.

Genes, of course, cannot be either selfish or unselfish; any more than they or any other non-conscious entities can engage in competition or make selections. (Natural selection is, notoriously, not selection; while it is a somewhat less familiar logical fact that, below the human level, the struggle for existence is not in the ordinary full sense competitive.) But this does not stop Dawkins proclaiming that his book 'is not science fiction; it is science . . . We are survival machines – robot vehicles blindly programmed to preserve the selfish molecules known as genes' (Dawkins, p. x). Although there are occasional disavowals later – 'like the paternosters of Mafia agents' (Midgley, p. 449) – Dawkins gives no warning here against construing this attribution literally. Literally is indeed how it has to be construed if his main thesis is to warrant the sensational and demoralizing conclusion drawn:

> The argument of this book is that we, and all other animals, are machines created by our genes. Like successful Chicago gangsters, our genes have survived, in some cases for millions of years, in a highly competitive world. This entitles us to expect certain qualities in our genes. I shall argue that a predominant quality to be expected in our genes is a ruthless selfishness . . . (Dawkins, pp. 2 and 3)

If any of this were true, then it would not be a bit of use to go on, as Dawkins does, to preach: 'Let us try to teach generosity and altruism, because we are born selfish' (ibid., p. 3). No eloquence could move programmed robots. But in fact it is none of it either true or even faintly sensible. Genes, as we have seen, do not and cannot necessitate our conduct. Nor are they capable of the calculation and understanding required to plot a course of either ruthless selfishness or sacrificial compassion. Furthermore and finally, the fastidious language-user cannot overlook that, while selfishness is not a matter only of having and pursuing an interest but of doing so in ways which improperly sacrifice the interests of others, altruism is not a necessarily self-sacrificial concern for others; and, even where self-sacrifice is involved, it is not always a sacrifice of reproductive success.

IV

PROGRESS, SOCIAL DARWINISM AND AN EVOLUTIONARY PERSPECTIVE

> The growth of a large business is merely a survival of the fittest . . . The American Beauty rose can be produced in the splendour and fragrance which bring cheer to its beholder only by sacrificing the early buds which grow up around it. This is not an evil tendency . . . It is merely the working out of a law of nature and a law of God.
>
> J. D. Rockefeller, in a Sunday School address.

> The historical process is the organizer of the City of God . . . the curve of the development of human society pursues its way across the graph of history with statistical certainty . . . the new world-order of social justice and comradeship, the rational and classless world-state, is no wild idealistic dream, but a logical extrapolation from the whole course of evolution, having no less an authority than that behind it . . .
>
> Joseph Needham, *Time, the Refreshing River*, pp. 16 and 41.

This final chapter is for two reasons by far the shortest. The first and less weighty is that anyone who, after reaching the end, still wants more can be referred to an exhaustive and perhaps exhausting treatment of all the topics now to be discussed in my monograph, *Evolutionary Ethics* (Flew, 1967). The second reason is that all the preparatory work, work which consumes much space in that monograph, has been done here in the three previous chapters. Like Caesar's Gaul the present treatment is divided into three parts: the first asks whether the record of biological evolution in the past can provide any kind of guarantee of future progress; the second questions whether it is in principle possible validly to derive prescriptive conclusions from the neutrally descriptive premises of scientific discovery; while the third goes on to suggest that, although no such derivation is possible, we nevertheless may, and sometimes should, see our conduct within an evolutionary perspective.

1. A GUARANTEE OF PROGRESS?

(a) The first suggestion that Darwin's theory promises progress is in the subtitle of *The Origin of Species – or The Preservation of Favoured Races in the Struggle for Life* – and in Darwin's later use of that catchy but unfortunate phrase 'survival of the fittest'. This phrase he appears to have borrowed from Herbert Spencer, an author for whose philosophical works Darwin in his own later years expressed vast respect. Yet, as has by now so frequently been pointed out, the survival of the fittest is here guaranteed only and precisely in so far as actual survival is the criterion of fitness to survive. In effect in this context 'fitness' is defined as 'having whatever it may as a matter of fact take to survive'. It is notorious that such biological fitness may be by other and independent standards most unadmirable. An individual or a species can have many splendid physical or mental endowments without these being or ensuring what is in fact needed for survival. Men who are in every way wretched creatures may, and all too often do, kill superb animals; while genius has frequently been laid low by the activities of unicellular beings having no wits at all.

It is in this light that we have to discount one or two over-optimistic mis-statements of the implications of the theory made by Darwin himself. The chapter 'Instinct', for instance, ends with the sentence: 'Finally . . . it is far more satisfactory to look at such instincts as the young cuckoo ejecting its foster-brothers, . . . as small consequences of one general law leading to the advancement of all organic beings, namely, multiply, vary, let the strongest live and the weakest die' (Darwin, C., 1859, p. 263).

But it simply does not follow that all organic beings must be in some line of progress. Innumerable lines in fact become extinct; and, as the World Wildlife Fund so regularly and so rightly reminds us, extinction is for ever. Darwin was, of course, perfectly well aware of this; as comes out rather nicely in a passage which also manifests his strong feeling for the importance of falsifiability. In a letter to Hooker, dated 18 April 1860, Darwin tells of an able but hostile review: 'But one argument is funny. The reviewer says, that if the doctrine were true, geological strata would be full of monsters which have failed' (Darwin, F., II p. 304).

(b) Second, traditional contrasts between higher and lower animals, combined with the more recent recognition that the former are all among the later products of the evolutionary process, may raise hopes of finding in living nature a passable substitute for Matthew Arnold's God: 'something, not ourselves, which makes for righteousness'. In his famous early essay on 'Progress, Biological and Other' Julian Huxley, launching what became a lifelong quest for such a substitute, contrived to detect several trends, which he felt able to commend as progressive, in the evolutionary story as that has so far become available. The same essay took as one of its mottoes the final sentence in the penultimate paragraph of *The Origin of Species*: 'As all the living forms of life are the lineal descendants of those which lived long before the Cambrian epoch, we may feel certain that . . . no cataclysm has desolated the whole world. Hence we may look with some confidence to a secure future of great length. And as natural selection works solely by and for the good of each being, all corporeal and mental endowments will tend to progress towards perfection' (Darwin, 1859, p. 459).

In this passage, too, the conclusions drawn in the final sentence do not in fact follow from the theory. The further reason why has a wider significance worth noting. Gould (1977) contains an excellent statement of the crux: 'No matter how much power Dawkins wishes to assign to genes, there is one thing he cannot give to them – direct visibility to natural selection. Selection simply cannot see genes and pick among them directly. It must use bodies as an intermediary . . . It accepts or rejects entire organisms . . .' This proposition totally discredits the supposed scoop revelation of gangster genes plotting 'with ruthless selfishness' their own individual survival. But it also indicates that any improvement in any single organ or function may at any time be enveloped in some general catastrophe for the whole organism consequent upon some fatal weakness elsewhere.

This once said, it has also to be insisted that to serve Huxley's purpose trends are useless. For mere trends may at any time stop, or go into reverse. What Huxley needs is at the very least a law of tendency or, better, an historicist law of development. Darwinism, despite two or three misleading hints, offers no basis for either. We know that Darwin pinned into his copy of the *Vestiges* a memorandum slip: 'Never use the words "higher" and "lower"' (Darwin and Seward, I p. 114n.). In the peroratory paragraph of *The Origin of Species* this self-denying ordinance is forgotten: '. . . from the war of nature, from famine and death, the most exalted object which we are capable of conceiving, namely the production of the higher animals, directly follows. There is a grandeur in this view of life . . .' (Darwin, 1859, p. 459).

There is indeed 'a grandeur in this view of life', and it is something which should be part of the world-outlook of every modern person. But what certainly is not to be found in Darwin's theory is the guarantee of progress for which Huxley and others have sought so long and so pathetically. Huxley craved, as he wrote in that first book of essays, 'to discover something, some being or power, some force or tendency . . . moulding the destinies of the world – something not himself, greater than himself, with which [he could] harmonize his nature . . . repose his doubts . . . achieve confidence and hope' (Huxley, J., 1923, p. 17).

(c) The third and most fundamental reason why no such guarantee can be found either in Darwinism or in any other theory of subhuman evolution emerges as we ponder what was said in Section 1 of Chapter III, in discussing the relations and lack of relations between the conceptual frameworks of Malthus and of Darwin. The crux is: that man is the creature which can, and cannot but, make choices; and that it is upon the senses of these choices that the future not only of mankind but also of almost all other species now largely depends. So, if there is any legitimate satisfaction anywhere for the secular religious cravings of Huxley and his like, then it can only be in the distinctively human sciences. Such persons should, and no doubt did, resonate to the later repudiated words of a marxizing British poet of the 1930s:

> Our day is our loss, Oh show us
> History the operator, the
> Organizer, Time the refreshing river
>
> W. H. Auden, 'Spain'

The soundness of this third and most fundamental objection comes out most harshly when we notice why we cannot echo Darwin's conclusion: 'Hence we may look with some confidence to a secure future of great length.' Armageddon apart, Quinton was stating the simple truth when he said: '[Man] has certainly won the contest between animal species in that it is only on his sufferance that any other species exist at all, amongst species large enough to be seen at any rate' (Quinton, p. 120). It should, therefore, be no wonder that the biochemist author of the second of this chapter's motto quotations was, as apparently he still is, a Marxist; whose assurance of victory was, presumably, based upon the historicist prophecies of Marx rather than derived from a direct and independent analysis of the implications of *The Origin of Species*. Certainly Needham followed *Time, the Refreshing River* with *History is on Our Side*.

2. FROM *'IS'* TO *'OUGHT'*

(a) The variety of policy morals which have in fact been drawn from Darwinism, whether validly or invalidly, is vast. There are also, as we shall be seeing, differences in the drawing: it may be a matter of indicating what is supposed to be an entailment; or it may be one of suggesting other and logically weaker kinds of connection. Let us now add to the first motto two other examples of recommended norms of human conduct being presented as pretty direct deductions from evolutionary biology. Since often nowadays people who employ the expression 'Social Darwinism' intend it to imply a commitment to what they believe to have been the political economy of Herbert Spencer, it is worth noting as we review all these three examples that other and incompatible systems of norms would appear to have an equally good claim to that title.

First, then, we can hear Adolf Hitler saying:

> If we did not respect the law of nature, imposing our will by the right of the stronger, a day would come when the wild animals would again devour us – then the insects would eat the wild animals, and finally nothing would exist upon earth except the

microbes . . . By means of the struggle the élites are continually renewed. The law of selection justifies this incessant struggle by allowing the survival of the fittest. Christianity is a rebellion against natural law, a protest against nature. Taken to its logical extreme Christianity would mean the systematic cult of human failure. (quoted Trevor-Roper, pp. 39 and 51; compare Bullock, pp. 36, 89, 398–9, 672 and 693)

Consider next, as a third example, something written in 1905, at a time when Hitler was taking lessons in the political exploitation of anti-semitism from Karl Lueger, the Christian Socialist Burgomeister of Vienna. James Ramsay Macdonald, who was to become the first Labour Party Prime Minister of Britain, wrote in a Preface to Enrico Ferri's *Socialism and Positive Science* that 'the conservative and aristocratic interests in Europe have armed themselves for defensive and offensive purposes with the law of the struggle for existence, and its corollary, the survival of the fittest. Ferri's aim in this volume has been to show that Darwinism is not only not in intellectual opposition to socialism, but is its scientific foundation.' Macdonald goes on to conclude that 'Socialism is naught but Darwinism economized, made definite, become an intellectual policy, applied to the conditions of human society' (Ferri, pp. v and vi–vii).

(b) The three examples given in the previous subsections are all examples of committing what is now labelled the Naturalistic Fallacy. They all, that is to say, contain or presuppose attempts to deduce normative prescriptions or proscriptions from premises containing only statements of neutral and uncommitted fact. Those who in recent years have taken to denying that the Naturalistic Fallacy is indeed a fallacy may be invited to explain: either what is the alternative basis of the invalidity of these particular arguments; or else how three valid arguments from the same premises come to yield mutually incompatible practical conclusions. For discussion both of this substantive question and of the hermeneutic issue of whether Hume was in one of his most quoted paragraphs intending to represent the attempt to deduce an *ought* from an *is* as fallacious (Hume 1739–40, III i 1, pp. 469–70), see Hudson.

(i) About these factitiously contentious matters I will here treat myself to one sharp, direct observation. It is that those who want to answer either or both questions in the negative are often almost unbelievably inept and naïve. Benjamin Gibbs, for instance, in an authoritarian book unpersuasively pretending not to be 'a tract *against* freedom', fails to catch Hume's irony. So he thinks to dispose of him as a spokesperson for the false and foolish thesis that all utterances containing the copula 'is' are straightforwardly descriptive, whereas all utterances with the copula 'ought' are purely prescriptive or proscriptive (Gibbs, pp. 8 and 116). Yet the whole point with both Hume's Law and Hume's Fork is precisely not that the crucial distinction always is made, but that it always should be (Flew, 1979, pp. 27–8 and 112–13). 'Hume's Law', it has to be said, is a nickname for the insistence that an 'ought' cannot be validly deduced from an 'is'; whereas 'Hume's Fork' describes the aggressive employment of the categorial distinction explained in Chapter II, 1 (a) (ii), above.

(ii) Moving from the general to the particular, we can now point out that we cannot validly infer, what Adolf Hitler would clearly have wished us to infer, that the groups which survive in such a struggle for existence must also be rated élites by some other and more exacting criterion than that of mere survival. It is now time to insist, as was not done before, that for evolutionary theory the crux is neither mere survival, nor survival at the top of some heap, but survival to reproduce; and hence the maintenance, sometimes the multiplication of the species. Many sincerely professing Social Darwinists have been so misguided as to select for their commendation, as 'survivors indicated in the struggle for existence', putative élites which have in fact produced and raised to maturity fewer offspring than the vulgar losers with whom they are thus to be compared. Hitler himself, so far as is known, had no children at all; the largest family among his closest associates, that of Goebbels, destroyed itself in the *Götterdämmerung* of 1945; while the 'master race' under their leadership suffered no population explosion on the scale common in this century among the 'inferior races' of Asia, Africa and Latin America.

(iii) Another key expression here is 'natural law'. A descriptive law of nature contains as part of its meaning that whatever is determined by its terms must be as a matter of fact necessary while whatever is incompatible with them must as a matter of fact be impossible. It is, by contrast, essential to the meaning of a prescriptive or a proscriptive law of nature that whatever is determined by its terms should be enjoined or, as the case may be, forbidden. Although it is not, I suppose, essential to the second sort of law that there should be alternative courses open to those to whom it is addressed, as it is to the first sort that there must within its scope be no alternative possibilities, still the proclamation of a prescriptive or a proscriptive law loses all point if it is the case that to a person its public will in fact either 'obey', or 'disobey' – whichever it is – necessarily and willy-nilly.

Consider for a moment the to us surprising definition of 'law of nature' given by the most formidable of British political thinkers in Chapter XIII of *Leviathan*. Hobbes says there that a law of nature is 'a precept or general rule, found out by reason, by which a man is forbidden to do that, which is destructive of his life or taketh away the means of preserving the same; and to omit that, by which he thinketh it may best be preserved'.

Compare now this prescriptive and proscriptive formulation with his statement in Chapter I of *De Cive*:

> Among so many dangers therefore as the natural lusts of men do daily threaten each other withal, to have a care of one's self is not a matter to be so scornfully looked upon, as if so there had been a power and will left in one to have done otherwise. For every man is desirous of what is good for him, and shuns what is evil, but chiefly the chiefest of natural evils, which is death; and this he doth, by a certain impulsion of nature, no less than that whereby a stone moves downwards.

It is remarkable that so incisive a thinker and so excellent a writer does not here make it absolutely clear whether what it is that we are supposed not to be able to avoid is having the desires (which is true) or acting to satisfy them (which is false). In so far as Hobbes was one of the first of many to aspire to develop a psychological mechanics, in which our desires (or 'drives') would be flagrantly misconstrued as forces compelling us to act towards their satisfaction willy-nilly, we may say that he slips into such ambiguity *quasi veritate coactus* (as if compelled by the truth) (Flew, 1978, VII and passim).

It must be, therefore, in the light of all this, egregiously perverse to appeal to some Hobbist supposed psychological Law of Self-preservation in order to justify a shabby skulking out of danger; perhaps adding, with a touch of sanctimoniousness, that this misconduct is categorically imperative under the truly Hobbist prescriptive Law of Self-preservation. Your appeal to the second (prescriptive) law constitutes a tacit admission that the supposed first (descriptive) law does not in fact hold. To grasp that and why this manoeuvre is so perverse and so preposterous is at the same time to grasp that and why the Naturalistic Fallacy is, and cannot but be, fallacious.

Prescriptions and proscriptions refer to choice. They refer, that is, to those areas of human behaviour where alternatives are open to the behaver; who is thus and there truly an agent. But choice is something which, deliberately and with good reason, Darwin leaves out from *The Origin of Species*; as he leaves out everything else in any way distinctive of life above the subhuman level. It is with this intended omission in mind that, as we saw in Section 1 of Chapter III, Darwin stipulates that his fundamental principle of population must exclude our own species from its scope. Furthermore, as was also brought out there, his natural selection is precisely not selection. It should be seen as strange and incongruous that Darwin's theory has so often been seen as providing imperative direction for what he himself was here at such pains not to touch.

(iv) Wherever such inferences are made to normative conclusions supposedly entailed by what is in fact a purely descriptive and explanatory theory, the chances are that the term 'natural' will play some mediating part. The context of Hume's classical description of the nerve of that invalid form of argument, for which G. E. Moore was – nearly two centuries later – to introduce the apt label 'Naturalist Fallacy', is here significant. This description constitutes the final, ostensibly afterthought paragraph of the first of the two sections of Part I of Book III of the *Treatise*. Perhaps the most powerful reason for dismissing recent suggestions that Hume did not intend this luminous observation to be read in the way in which we have been reading it is the fact that in the second section of Part II he proceeds to argue, with several illustrations, 'that nothing can be more unphilosophical than those systems, which assert, that virtue is the same with what is natural and vice with what is unnatural' (Hume, 1739–40, III ii 2 p. 475).

Hume's reproach goes home against the first two 'systems' already cited by us. For Hitler assails Christianity as 'rebellion against natural law, a protest against nature', while Rockefeller, addressing a Christian Sunday School, takes the same wrong way to come up with in part an opposite conclusion: '. . . survival of the fittest . . . is merely the working out of a law of nature and a law of God' (Ghent, p. 29). It is also just worth noting Rockefeller's inept choice of examples: as, surely, an achievement of the nurserypersons the American Beauty rose is precisely not a trophy of Darwinian natural selection; and so it belongs here, if at all, as a specimen of the artificial rather than of the natural.

3. SEEING IN AN EVOLUTIONARY PERSPECTIVE

The upshot of Section 2, immediately above, is, surely, that any attempt to deduce norms for human conduct from this theory of the origin of species by natural selection must be as irredeemably wrong-headed as the search in the same area for some guarantee of human progress, some assurance of victory for the right; or, if you insist, the left. So far, so negative; and, perhaps, so dispiriting. There is, however, a third way in which evolutionary biology can bear on such practical issues. We can and, I believe, at least occasionally we should, try to see our human activities and inactivities in an evolutionary perspective.

(a) Julian Huxley provides an excellent illustration. In his younger days, as we saw earlier, he had hoped to be able to squeeze out something stronger than simply seeing in a perspective. But later he seemed to realize that this could not be done. Instead in his latest books he urged that the correct way for the moralist and the politician to take account of Darwinian discoveries was quite different. Thus, in the Preface to *Evolution in Action*, Huxley maintained:

> It makes a great difference whether we think of the history of mankind as wholly apart from the rest of life, or as a continuation of the general evolutionary process, though with special characteristics of its own . . . In the light of evolutionary biology, man can now see himself as the sole agent for further evolutionary advance on this planet . . . He finds himself in the unexpected position of business manager for the cosmic process of evolution. (Huxley, 1953, pp. vii and 132)

The questions now arise: both whether it does indeed make a great difference; and whether such a way of seeing the human situation has any claims on anyone who has not enjoyed a biological training. What is to be said against the impatience of Jane Welsh Carlyle? Mrs Carlyle, it may be recalled, confessed that she 'did not feel that the slightest light could be thrown on my practical life for me, by my having it ever so logically made out that my first ancestor, millions of millions of ages back had been, or even had not been, an oyster' (Carlyle, III pp. 20–1).

The first point to make in reply is that the evolutionary perspective takes for granted various general propositions which are in fact true. Their truth often bears on that of others which do in fact play crucial parts in various conduct-guiding world-outlooks. We have in Chapter II considered some of these implications at length. The second and third points both constitute familiar justifications for 'taking a wider view', though they are not on that account to be dismissed as inconsiderable. One is that such wider views may enable us to see things which do not emerge so easily, if at all, from a more blinkered appreciation. This point can be well illustrated by the case of Julian Huxley himself. For it was precisely an evolutionary vision which determined his own recognition, long locust years before this was even as widely admitted as it is now, that drastic checks on human fertility are a necessary condition for the maintenance of man's estate – to say nothing of its relief. It was the same vision which, two generations earlier, had determined his grandfather's recognition that population problems must be of paramount importance (Huxley, T. H., 1890, pp. 315–16).

The other remaining point is that many people long to see things as a whole, to find some deep, comprehensive, world-picture against which they may set their lives. No philosopher has any business either to despise or not to share such yearnings. Here the Darwinian vision possesses the neither universal nor despicable merit of being founded upon, and being not incompatible with, known facts. Also, as Darwin said in his peroration: 'There is a grandeur in this view of life . . .' (Darwin, C. 1859, pp. 459–60).

(b) The possible importance of having the kind of perspective which Julian Huxley had comes out most sharply through a contrast. If we were to judge only by his published writings we should have to put that philosophers' philosopher G. E. Moore down as in his interests passionately parochial and blinkered. In that most curious volume, *Principia Ethical*, for instance, the whole discussion both of ethics and of meta-ethics proceeds as if outside time and space: Moore and his Cambridge and Bloomsbury culture-circle appear never to have heard travellers or historians tell of people cherishing very different values. The argument is equally undisturbed by any news from the science front: Moore might as well have been writing not merely before Darwin but before Newton.

Take, for instance, Moore's treatment of the Naturalistic Fallacy, and compare it with that of Hume; whose name, incidentally, scores no mention in Moore's book. Had Moore seen man as a part of nature, and in an evolutionary perspective, how could he have gone on as he did about the detection of his strange, inexplicably non-natural characteristics? How could he then have failed to conclude, as, without benefit of Darwin, Hume had concluded nearly two centuries earlier, that value judgements, which are not deducible from anything neutrally referring only to what is the case, must instead involve some kind of projection of individual or collective human activity or desire; much as Newton and Galileo had concluded that the so-called secondary qualities are projections on to 'things in themselves' of what is really only 'in the mind': the only authentic qualities of those things in themselves being the primary? (Flew, 1979, pp. 42–3 and 69–72)

(c) I can think of no better way of finishing this entire book than by repeating the conclusion of that earlier monograph, *Evolutionary Ethics* (Flew, 1967, p. 60). Having there as previously here quoted part of a purple patch from *Evolution in Action* I concluded by repeating that quotation and giving a bit more:

> In the light of evolutionary biology man can now see himself as the sole agent of further evolutionary advance on this planet, . . . He finds himself in the unexpected position of business manager for the cosmic process of evolution. He no longer ought to feel separated from the rest of nature, for he is part of it – that part which has become conscious, capable of love and understanding and aspiration. He need no longer regard himself as insignificant in relation to the cosmos. (Huxley, J., 1953, p. 132)

BIBLIOGRAPHY

The list which follows is intended to include all works mentioned in the text, providing the bibliographical particulars necessary for pursuing references. In some cases, where the primary source is hard to track down, suggestions have been given of more accessible secondary sources.

Before proceeding to that long list, a few suggestions for possible further reading may be welcome, notwithstanding that most of these have in effect been made in the text above. At this stage it would be the sheerest self-indulgence to take up Moorehead's *Darwin and the Beagle*. But it could not fail to be a most agreeable self-indulgence, thanks especially to the nostalgic and carefully chosen illustrations. The urgent call must be to read something by Darwin himself, the prime claimants being *The Voyage of the 'Beagle'* and *The Origin of Species* itself. The first of these was in the so often and so unfairly despised Victorian Age a best-seller, and it seems never since to have gone out of print. The latter is beyond question the most important of all those rather few classics of the natural sciences which are immediately accessible to the inexpert general reader.

The best biography is still that by Gavin de Beer (*Charles Darwin: A Scientific Biography*); while Darwin's characteristically open and engaging *Autobiography* must, for reasons already given, be read only in the full text presented by his granddaughter Nora Barlow. For more on the case for Darwinism and Neo-Darwinism, see Michael Ruse *Darwinism Defended*, and further recommendations there. For the background of the previous revolution in geology, see C. C. Gillispie *Genesis and Geology*. For a similar history of ideas treatment of *The Darwinian Revolution*, try Michael Ruse's book under that title; but, on no account, Himmelfarb. For further reading on more particular topics discussed in the text, the best course is to follow up some of the references given there: on sociobiology, for instance, Michael Ruse *Sociobiology: Sense or Nonsense?*; and, on the putative moral implications of Darwinism, my own monograph *Evolutionary Ethics*.

BIBLIOGRAPHY OF BOOKS AND ARTICLES TO WHICH REFERENCE HAS BEEN MADE

Abercrombie, J. (1838) *Inquiries concerning the Intellectual Powers* (London: Murray)

Aquinas, St T. (1963 onwards) *Summa Theologica*, new translation and edition by various hands under Thomas Gilby as general editor (London, and New York, Eyre and Spottiswoode, and McGraw-Hill)

Augustine, St of Hippo (1950) *On the Greatness of the Soul*, translated by J. M. Colleran (Westminster, Md., and London: Newman, and Longmans Green)

(1966) *The Catholic and Manichean Ways of Life*, translated by D. A. and I. J. Gallagher (Washington: Catholic UP)

Austin, J. L. (1961) 'A Plea for Excuses' in *Philosophical Papers* (Oxford: Clarendon)

Bakunin, M. (1974) *Statism and Anarchism*, translated by H. J. Frank, (Brooklyn, N.Y.: Revisionist)

Barlow, N. (ed.) (1968) *The Autobiography of Charles Darwin* (London: Collins)

Benda, J. (1969) *The Treason of the Intellectuals* (La Trahison des Clercs), translated by R. Aldington (New York: Norton)

Bernier, F. (1891) *Travels in the Mogul Empire: A.D. 1656–1668*, translated by I. Brock, revised by A. Constable (London: Constable)

Bullock, A. (1962) *Hitler: A Study in Tyranny* (Harmondsworth: Penguin Books)

Burnett, J. (Lord Monboddo) (1774) *The Origin and Progress of Language* (2nd edn. Edinburgh)

Carlyle, J. W. (1883) *Letters and Memorials* (London: Longmans Green)

Chambers, R. (1884) *Vestiges of the Natural History of Creation* (12th edn. London and Edinburgh: Chambers)

Cornforth, M. (1968) *The Open Philosophy and the Open Society* (London: Lawrence and Wishart)

Darwin, C. (1831–6) *Charles Darwin's Diary of the Voyage of HMS 'Beagle'*, edited by N. Barlow (Cambridge: CUP 1933)

(1839) *The Voyage of the 'Beagle'* (London, and New York: Dent, and Dutton, 1959)

(1842) *The Structure and Distribution of Coral Reefs: Being the First Part of the Geology of the Voyage of the 'Beagle'* (London)

(1859) *The Origin of Species*, edited by J. B. Burrow (Harmondsworth, and Baltimore: Penguin Books, 1968. This is a reprint of the 1859 1st edn.)

(1868) *Variation of Plants and Animals under Domestication* (London: Murray)

(1871) *The Descent of Man* (2nd edn. London: Murray, 1874)

(1872) *The Expression of the Emotions of Man and Animals* (London: Murray)

(1879) *The Life of Erasmus Darwin* (London: Murray)

(1881) *The Formation of Vegetable Mould through the Action of Worms* (London: Murray)

Darwin, E. (1794–7) *Zoönomia, or the Laws of Organic Life* (London)

Darwin, F. (ed.) (1887) *The Life and Letters of Charles Darwin* (London: Murray)

Darwin, F. and Seward, A. C. (eds) (1903) *More Letters of Charles Darwin* (London: Murray)

Dawkins, R. (1976) *The Selfish Gene* (New York: OUP)

de Beer, G. (1963) *Charles Darwin: A Scientific Biography* (London: Nelson)

Denzinger, H. (ed.) (1953) *Enchiridion Symbolorum* (Freiburg-im-Breisgau: Herder) (29th revised edn.)

Descartes, R. (1931) *The Philosophical Works of Descartes*, translated by E. S. Haldane and G. R. T. Ross (rev. edn. Cambridge: CUP)

Dobzhansky, T. (1951) *Genetics and the Origin of Species* (3rd edn. New York: Columbia UP)

(1963) 'Anthropology and the Natural Sciences – The Problem of Human Evolution' in *Current Anthropology*, IV 138

Durkheim, E. (1965) *Montesquieu and Rousseau, Precursors of Sociology* (Ann Arbor, Mich.: Michigan UP)

Eddy, M. B. G. (1875) *Science and Health* (Boston: First Church)

Engels, F. (1844) *Outline of a Critique of Political Economy*, in D. K. Struik (ed.) *The Economic and Philosophical Manuscripts of 1844 by Karl Marx* (New York: International, 1964)

(1878) *Socialism: Utopian and Scientific*, translated by E. Aveling (New York: Path, 1972)

Eysenck, H. J. (1975) *The Inequality of Man* (London: Collins Fontana)

Ferguson, A. (1767) *An Essay on the History of Civil Society*, edited by D. Forbes (Edinburgh: Edinburgh UP, 1966)

(1792) *Principles of Moral and Political Science* (London and Edinburgh)

Ferri, E. (1906) *Socialism and Positive Science* (London: Independent Labour Party)

Fisher, H. A. L. (1933) *Our New Religion* (London: C. A. Watts)

Flew, A. G. N. (1961) *Hume's Philosophy of Belief* (London: Routledge and Kegan Paul)

(ed.) (1964) *Body, Mind and Death* (New York, and London: Macmillan, and Collier-Macmillan)

(1966) *God and Philosophy* (London: Hutchinson). This is to be republished in 1984 under a fresh title by Open Court of La Salle, Ill.

(1967) *Evolutionary Ethics* (London: Macmillan). This monograph is also printed in W. D. Hudson (ed.) *New Studies in Ethics* (London: Macmillan, 1974), vol. II, pp. 217–86

(1970) 'Introduction' to *Malthus on Population* (Harmondsworth: Penguin Books). The text is that of the *First Essay*. The *Second Essay* is given in the current Everyman edition

(1971) *An Introduction to Western Philosophy* (London, and Indianapolis: Thames and Hudson, and Bobbs-Merrill)

(1975) *Thinking about Thinking* (London: Collins Fontana). Also from Prometheus Books of Buffalo as *Thinking Straight*

(1976a) *Sociology, Equality and Education* (London, and New York: Macmillan, and Barnes and Noble)

(1976b) *The Presumption of Atheism* (London: Pemberton/Elek). This is to be republished in 1984 by Prometheus Books of Buffalo, under a different title

(1978) *A Rational Animal* (Oxford: Clarendon)

(1979) *Philosophy: An Introduction* (London: Hodder and Stoughton). This is in the Teach Yourself series.

(1982a) 'Finch on Wittgenstein and Radnitzky on Popper' in *Proceedings of the Xth International Congress on the Unity of the Sciences (Seoul) 1981* (New York: International Cultural Foundation)

(1982b) 'Another Idea of Necessary Connection' in *Philosophy* LVI 222

(1983a) 'Human Choice and Historical Inevitability' in *Journal of Libertarian Studies* V 4 (1983, though falsely dated 1981)

(1983b) 'A Strong Programme for the Sociology of Belief' in *Inquiry* XXV

(1984?) 'Prophecy or Philosophy, Historicism or History?' in C. Wilson and R. Duncan *Marx Refuted* (Oxford: Pergamon, forthcoming)

Flew, A. G. N. and MacIntyre, A. C. (eds) (1955) *New Essays in Philosophical Theology* (London: SCM Press)

Flew, A. G. N. and Warren, T. B. (1977) *The Warren–Flew Debate* (Jonesboro, Arkansas: National Christian Press)

Freud, S. (1922) *Introductory Lectures on Psycho-Analysis*, translated by Joan Rivière (London: Allen and Unwin)

(1924–50) *Collected Papers*, translated by Joan Rivière and others (London: Hogarth)

Ghent, W. J. (1902) *Our Benevolent Feudalism* (New York: Macmillan)

Gibbs, B. (1976) *Freedom and Liberation* (Brighton: Harvester). Now from the Sussex University Press.

Gillispie, C. C. (1951) *Genesis and Geology* (Cambridge, Mass.: Harvard UP). Now a Harper Torchbook

Gish, D. T. (1973) *Evolution: The Fossils Say No!* (New York: Clear Light Publications)

Gosse, E. (1959) *Father and Son* (London: Four Square-Landsborough)

Gosse, P. H. (1873) *The Natural History of Creation* (London: van Vorst)

Gould, S. J. (1977) 'Caring Groups and Selfish Genes' in *Natural History* LXXXVI (December)

(1978) *Ever since Darwin* (London: Burnett and Deutsch)

Gould, S. J. and Eldredge, N. (1977) 'Punctuated Equilibria: the Tempo and Mode of Evolution Reconsidered' in *Paleobiology* III 3 (Spring)

Gruber, H. E. (1974) *Darwin on Man* (London: Wildwood House)

Haldane, J. B. S. (1937) *The Inequality of Man* (Harmondsworth: Penguin)

(1954) 'The States of Evolution' in *Evolution as a Process* (London)

Hayek, F. A. (1960) *The Constitution of Liberty* (London: Routledge and Kegan Paul)

(1967) 'Results of Human Action but not of Human Design' in his *Studies of Philosophy, Politics and Economics* (London: Routledge and Kegan Paul)

Herschel, J. F. W. (1830) *Preliminary Discourse to the Study of Natural Philosophy* (London: Longmans Brown)

Himmelfarb, G. (1959) *Darwin and the Darwinian Revolution* (London: Chatto and Windus)

Hobbes, T. (1642) *De Cive or The Citizen*, edited by S. P. Lamprecht (New York: Appleton-Century-Crofts, 1949)

Hook, S. (1943) *The Hero in History* (New York: Humanities)

Hudson, W. D. (ed.) (1969) *The Is/Ought Question* (London: Macmillan)

Humboldt, A. von (1812–29) *Personal Narrative of Travels: 1799–1804* (London: Longman, Hurst)

Hume, D. (1739–40) *A Treatise of Human Nature*, edited by L. A. Selby-Bigge, revised by P. H. Nidditch (Oxford: Clarendon, 1976)

(1748) *An Inquiry concerning Human Understanding* – the *First Inquiry*, in *Enquiries*, edited by L. A. Selby-Bigge, revised by P. H. Nidditch (Oxford: Clarendon, 1975)

(1752) *An Inquiry concerning the Principles of Morals*, included in the above – the *Second Inquiry*

(1757) *The Natural History of Religion*, edited by H. E. Root (London: A. and C. Black, 1956)

(1779) *Dialogues concerning Natural Religion*, edited by N. Kemp Smith (Edinburgh: Nelson, 1947). This standard edition has since been reprinted in at least two different paginations

Hutton, J. (1795) *Theory of the Earth, with Proofs and Illustrations* (Edinburgh; Creech)

Huxley, J. (1923) *Essays of a Biologist* (Harmondsworth: Penguin Books)

(1953) *Evolution in Action* (London: Chatto and Windus)

Huxley, L. (ed.) (1900) *Life and Letters of Thomas Henry Huxley* (London: Macmillan)

Huxley, T. H. (1859) 'The Darwinian Hypothesis' reprinted in T. H. Huxley *Man's Place in Nature* (London, and New York: Dent, and Dutton, 1906)

(1859) *The Times* review of *The Origin of Species* reprinted in *Man's Place in Nature*

(1860) 'Darwin on the Origin of Species' in the *Westminster Review* for 1860, reprinted in *Man's Place in Nature*

(1887) 'The Progress of Science' in T. H. Huxley *Method and Results* (London: Macmillan, 1894)

(1890) 'On the Natural Inequality of Man', in his *Method and Results* (London: Macmillan, 1894)

(1906) 'A Critical Examination of the Position of Mr Darwin's work On the *Origin of Species*' reprinted in *Man's Place in Nature*

Jones, R. (1831) *An Essay on the Distribution of Wealth and the Sources of Taxation* (London)

Kettlewell, H. B. D. 'Selection Experiments on Industrial Melanism in the Lepidoptera' in *Heredity* IX, pp. 323–42

Keynes, J. M. (1921) *A Treatise on Probability* (Cambridge: CUP)

Kirk, G. S. and Raven, J. E. (1957) *The Pre-Socratic Philosophers* (Cambridge: CUP)

Kolakowski, L. (1978) *Main Currents of Marxism* (Oxford: Clarendon)

Koyré, A. (1958) *From the Closed World to the Infinite Universe* (New York: Harper and Row)

Kropotkin, P. (1902) *Mutual Aid: a Factor of Evolution* (London: Heinemann)

Lamarck, J.-B. de (1963) *Zoological Philosophy*, translated by H. Elliot (New York: Hafner)

Lasky, M. (1976) *Utopia and Revolution* (London: Macmillan)

Leibniz, G. W. F. (1951) 'Principles of Nature and of Grace' in P. P. Weiner (ed.) *Leibniz Selections* (New York: Scribner)

Lenin, V. I. (1970) *What the 'Friends of the People' are* (Moscow: FLPH)

Locke, J. (1975) *An Essay Concerning Human Understanding*, edited by P. H. Nidditch (Oxford: Clarendon)

Lovejoy, A. O. (1904) 'Some Eighteenth Century Evolutionists' in *Popular Science Monthly* (New York)

(1936) *The Great Chain of Being* (New York: Harper)

Lucretius (T. Lucretius Carus) *de Rerum Natura* (On the Nature of Things), translated by W. H. D. Rouse (Loeb Classical Library: London, and Cambridge, Mass.: Heinemann, and Harvard UP)

Lyell, C. (1830) *Principles of Geology* (London: Murray)

Magee, B. (1973) *Popper* (London: Collins Fontana)

Malthus, T. R. (1798) *An Essay on the Principle of Population*, edited by A. Flew (Harmondsworth and Baltimore: Penguin Books, 1970) – the *First Essay*

(1802) *An Essay on the Principle of Population* (6th edn. London, 1826) – the *Second Essay*

(1824) *A Summary View of the Principle of Population*, in D. V. Glass (ed.) *An Introduction to Malthus* (London: Watts, 1953)

Marx, K. (1867) *Capital: A Critical Analysis of Capitalist Production*, translated by S. Moore and E. Aveling (Moscow: FLPH, 1961)

(1876) 'On Bakunin's *Statism and Anarchy*' in D. McClellan (ed.) *Karl Marx: Selected Writings* (Oxford: OUP, 1977)

Marx, K. and Engels, F. (1844) *The Economic and Philosophical Manuscripts of 1844 by Karl Marx*, edited by D. K. Struik (New York: International, 1944)

(1848) *The Communist Manifesto*, translated by S. Moore, edited by A. J. P. Taylor (Harmondsworth: Penguin Books, 1968)

Meek, R. C. (1953) The quoted passage, and many similar

explosions against Malthus, are most easily found in R. C. Meek *Marx and Engels on Malthus* (London: Lawrence and Wishart)

Midgley, M. (1979) 'Gene-juggling' in *Philosophy* LIV 210 (October)

Mill, J. (1820) *History of British India* (2nd edn. London)

Mill, J. S. (1848) *Principles of Political Economy* (Toronto: Toronto UP, 1965)

Moore, G. E. (1903) *Principia Ethica* (Cambridge: CUP)

Moorehead, A. (1969) *Darwin and the Beagle* (Harmondsworth and New York: Penguin Books)

Morris, D. (1967) *The Naked Ape* (London: Corgi)

(1970) *The Human Zoo* (London: Cape)

Needham, J. (1943) *Time, the Refreshing River* (London: Allen and Unwin)

(1946) *History is on Our Side* (London: Allen and Unwin)

Newton, I. (1962) *Mathematical Principles of Natural Philosophy*, translated by A. Motte, revised and edited by F. Cajori (Berkeley and Los Angeles: California UP)

Oldroyd, D. R. (1980) *Darwinian Impacts* (Milton Keynes: Open UP)

Page, L. R. (1983) *Marx and Darwin: The Unveiling of a Myth* (Beckenham, Kent: Centre for Liberal Studies)

Paley, W. (1838) *The Works of Dr. William Paley*, edited by E. Paley (London: Longmans)

Pantin, C. F. A. (1953) *A History of Science* (London: Cohen and West)

Popper, K. R. (1957) *The Poverty of Historicism* (London: Routledge and Kegan Paul)

(1959) *The Logic of Scientific Discovery* (London: Hutchinson)

(1963) *Conjectures and Refutations* (London: Routledge and Kegan Paul)

(1974) *An Intellectual Autobiography*, in P. A. Schilpp (ed.) *The Philosophy of K. R. Popper* (La Salle, Ill.: Open Court). This was later published separately as *Unended Quest*

(1978) 'Natural Selection and the Emergence of Mind' in *Dialectica* vol. 32 Fasc. 3–4. Compare also letters in *New Scientist* LXXXVII p. 611 (1980) and in *Science* CCXII, pp. 281 and 873 (1981)

(1979) *Objective Knowledge* (rev. edn. Oxford: Clarendon)
Prescott, W. H. (1936) *History of the Conquest of Mexico* and *History of the Conquest of Peru* (New York: Modern Library)
Quinton, A. M. (1966) 'Ethics and the Theory of Evolution' in I. T. Ramsey (ed.) *Biology and Personality* (Oxford: Blackwell)
Romer, A. S. (1958) 'Darwin and the Fossil Record' in S. A. Barnett (ed.) *A Century of Darwin* (London: Heinemann)
Royce, J. (1892) *The Spirit of Modern Philosophy* (New York: Houghton Mifflin)
Ruse, M. (1973) *The Philosophy of Biology* (London: Hutchinson)
(1979a) *Sociobiology: Sense or Nonsense?* (Dordrecht: Reidel)
(1979b) *The Darwinian Revolution* (Chicago: Chicago UP)
(1982) *Darwinism Defended* (Reading, Mass.: Addison-Wesley)
Ryle, G. (1949) *The Concept of Mind* (London: Hutchinson)
Schwarzschild, L. (1948) *The Red Prussian* (London: Hamilton)
Schweber, S. S. (1978) 'The Genesis of Natural Selection – 1838: Some Further Insights' in *Bioscience* XXVIII (May) pp. 321–6
Shafarevich, I. (1980) *The Socialist Phenomenon* (New York: Harper and Row)
Sherrington, C. (1946) *Man on his Nature* (Cambridge: CUP)
Simpson, G. G. (1976) *The Meaning of Evolution* (New Haven: Yale UP)
Smith, A. (1976) *An Inquiry into the Nature and Causes of the Wealth of Nations*, edited by R. H. Campbell and A. S. Skinner (Oxford: Clarendon)
Toulmin, S. E. and Goodfield, J. (1965) *The Discovery of Time* (London: Hutchinson)
Trevor-Roper, H. R. (ed.) (1953) *Hitler's Table Talk* (London: Weidenfeld and Nicolson)
Vorzimmer, P. J. (1977) 'Darwin's Reading' in *Journal of the History of Biology* (Fall)
Wesson, R. G. (1976) *Why Marxism? The Continuing Success of a Failed Theory* (London: Temple Smith)

Weyl, N. (1979) *Karl Marx: Racist* (New Rochelle, N.Y.: Arlington House)

Whately, R. (1832) *Lectures on Political Economy* (London)

Wightman, W. P. D. (ed.) (1980) *Adam Smith: Essays on Philosophical Subjects* (Oxford: Clarendon)

Wilson, E. O. (1971) *The Insect Societies* (Cambridge, Mass.: Harvard UP)

(1975) *Sociobiology: The New Synthesis* (Cambridge, Mass.: Harvard UP)

Wittfogel, K. A. (1981) *Oriental Despotism: A Comparative Study of Total Power* First Vintage Edition with Author's new Preface (New York: Random House)

Wittgenstein, L. (1923) *Tractatus Logico-Philosophicus*, translated by C. K. Ogden (London: Routledge and Kegan Paul). The Authorized Version

Wolfe, B. D. (1967) *Marxism: 100 Years of a Doctrine* (London: Chapman and Hall)

INDEX